Biotechnology in Agriculture, Industry and Medicine

PATHOGEN DETECTION METHODS: BIOSENSOR DEVELOPMENT

BIOTECHNOLOGY IN AGRICULTURE, INDUSTRY AND MEDICINE

Agricultural Biotechnology: An Economic Perspective
Margriet F. Caswell; Keith O. Fuglie, Cassandra A. Klotz
2003. 1-59033-624-0

**Governing Risk in the 21st Century:
Lessons from the World of Biotechnology**
Peter W.B. Phillips et al.(Editors)
2006. 1-59454-818-8

Biotechnology and Industry
G. E. Zaikov (Editor)
2007. 1-59454-116-7

Research Progress in Biotechnology
G. E. Zaikov (Editor)
2008. 978-1-60456-000-8

Biotechnology and Bioengineering
William G. Flynne (Editor)
2008. 978-1-60456-067-1

Biotechnology: Research, Technology and Applications
Felix W. Richter (Editor)
2008. 978-1-60456-901-8

Biotechnology: Research, Technology and Applications
Felix W. Richter (Editor)
2008. 978-1-60876-369-6

Biotechnology, Biodegradation, Water and Foodstuffs
G.E. Zaikov and Larisa Petrivna Krylova (Editors)
2009. 978-1-60692-097-8

Industrial Biotechnology
Shara L. Aranoff, Daniel R. Pearson, Deanna Tanner Okun, Irving A. Williamson, Dean A. Pinkert, Robert A. Rogowsky and Karen Laney-Cummings
2009. 978-1-60692-256-9

Industrial Biotechnology and the U.S. Chemical and Biofuel Industries
James R. Thomas (Editor)
2009. 978-1-60741-899-3

Biosensors: Properties, Materials and Applications
Rafael Comeaux and Pablo Novotny (Editors)
2009. 978-1-60741-617-3

Industrial Biotechnology: Patenting Trends and Innovation
Katherine Linton, Philip Stone, Jeremy Wise, Alexander Bamiagis, Shannon Gaffney, Elizabeth Nesbitt, Matthew Potts, Robert Feinberg, Laura Polly, Sharon Greenfield, Monica Reed, Wanda Tolson and Karen Laney-Cummings
2009. 978-1-60741-032-4

Biochemical Engineering
Fabian E. Dumont and Jack A. Sacco (Editors)
2009. 978-1-60741-257-1

Medicinal Plants: Classification, Biosynthesis and Pharmacology
Alejandro Varela and Jasiah Ibañez (Editors)
2009. 978-1-60876-027-5

Perspectives on Lipase Enzyme Technology
J. Geraldine Sandana Mala and Satoru Takeuchi
2009. 978-1-60741-977-8

Technologies and Management for Sustainable Biosystems
Jaya Nair, Christine Furedy; Chanakya Hoysala and Horst Doelle (Editors)
2009. 978-1-60876-104-3

Biosensors: Properties, Materials and Applications
Rafael Comeaux and Pablo Novotny (Editors)
2009. 978-1-61668-181-4

**Cellulose: Structure and Properties,
Derivatives and Industrial Uses**
Arnaud Lejeune and Thibaut Deprez (Editors)
2010. 978-1-60876-388-7

**Synthetic and Integrative Biology:
Parts and Systems, Design Theory and Applications**
James T. Gevona (Editor)
2010. 978-1-60876-678-9

**Biometals: Molecular Structures,
Binding Properties and Applications**
Gaillard Blanc and Damien Moreau (Editors)
2010. 978-1-60876-852-3

**Carbohydrate Binding Modules:
Functions and Applications**
Susana Moreira and Miguel Gama
2010. 978-1-60876-979-7

Bioengineering: Principles, Methodologies and Applications
Audric Garcia and Ciel Durand (Editors)
2010. 978-1-60741-762-0

**Biotechnology in Medicine, Foodstuffs,
Biocatalysis, Environment and Biogeotechnology**
Sergey D. Varfolomeev, Gennady E. Zaikov and Larisa P. Krylova
2010. 978-1-60876-902-5

Strategic Alliances in Biotechnology and Pharmaceuticals
Hans Gottinger, Celia Umali and Frank Floether
2010. 978-1-60876-997-1

Use of Organosilanes in Biosensors
*V. Dugas, C. Demesmay, Y. Chevolot
and E. Souteyrand*
2010. 978-1-61668-029-9

Edible Polysaccharide Films and Coatings
Pau Talens, María José Fabra and Amparo Chiralt
2010. 978-1-61668-191-3

**Bioactive Oligosaccharides: Production, Biological
Functions and Potential Commercial Applications**
Aneli M. Barbosa, Robert F. H. Dekker and Ellen C. Giese
2010. 978-1-61668-149-4

Pathogen Detection Methods: Biosensor Development
Eva Baldrich and Cristina Garcia-Aljaro
2010. 978-1-61668-298-9

**Synthetic and Integrative Biology:
Parts and Systems, Design Theory and Applications**
James T. Gevona (Editor)
2010. 978-1-61668-347-4

Use of Organosilanes in Biosensors
V. Dugas, Demesmay, Y. Chevolot and E. Souteyrand
2010. 978-1-61668-073-3

Edible Polysaccharide Films and Coatings
Pau Talens, María José Fabra and Amparo Chiralt
2010. 978-1-61668-493-8

Pathogen Detection Methods: Biosensor Development
Eva Baldrich and Cristina Garcia-Aljaro
2010. 978-1-61668-699-4

Biotechnology in Agriculture, Industry and Medicine

PATHOGEN DETECTION METHODS: BIOSENSOR DEVELOPMENT

EVA BALDRICH AND CRISTINA GARCIA-ALJARO

Nova Science Publishers, Inc.
New York

Copyright © 2010 by Nova Science Publishers, Inc.

All rights reserved. No part of this book may be reproduced, stored in a retrieval system or transmitted in any form or by any means: electronic, electrostatic, magnetic, tape, mechanical photocopying, recording or otherwise without the written permission of the Publisher.

For permission to use material from this book please contact us:
Telephone 631-231-7269; Fax 631-231-8175
Web Site: http://www.novapublishers.com

NOTICE TO THE READER

The Publisher has taken reasonable care in the preparation of this book, but makes no expressed or implied warranty of any kind and assumes no responsibility for any errors or omissions. No liability is assumed for incidental or consequential damages in connection with or arising out of information contained in this book. The Publisher shall not be liable for any special, consequential, or exemplary damages resulting, in whole or in part, from the readers' use of, or reliance upon, this material.

Independent verification should be sought for any data, advice or recommendations contained in this book. In addition, no responsibility is assumed by the publisher for any injury and/or damage to persons or property arising from any methods, products, instructions, ideas or otherwise contained in this publication.

This publication is designed to provide accurate and authoritative information with regard to the subject matter covered herein. It is sold with the clear understanding that the Publisher is not engaged in rendering legal or any other professional services. If legal or any other expert assistance is required, the services of a competent person should be sought. FROM A DECLARATION OF PARTICIPANTS JOINTLY ADOPTED BY A COMMITTEE OF THE AMERICAN BAR ASSOCIATION AND A COMMITTEE OF PUBLISHERS.

LIBRARY OF CONGRESS CATALOGING-IN-PUBLICATION DATA

Available upon Request
ISBN: 978-1-61668-298-9

Published by Nova Science Publishers, Inc. ✢ New York

Contents

Preface		xi
List of Abbreviations		1
Chapter 1	Pathogens at the Human Environment	5
Chapter 2	Conventional and Rapid Methods for Pathogen Detection	15
Chapter 3	Biosensors: An Alternative to Traditional Methods	23
Chapter 4	Biosensors for Pathogen Detection	81
Chapter 5	Drawbacks and Future Trends of Pathogen Biosensors	123
Index		127

PREFACE

Timely pathogen detection is an important issue for the efficient prevention of outbreaks in the populations worldwide. In this context, improvement of the existing methodology is thus directly applicable to a variety of fields, including clinical microbiology, food industry and environmental monitoring. The development of the sensor biotechnology in the last decades has set the basis for improved, faster and simplified detection of multiple analytes in complex matrices and a number of electrochemical, piezoelectric, and optical biosensors have been applied with different success to the detection of either whole bacteria or bacterial components. The main advantages of this new technologic approach rely on the rapid response, cost-effectiveness, easiness of manipulation and the possibility of performing *on-site* and *real-time* analysis of the samples compared to the traditional microbiology-based detection strategies for pathogen detection. In this booklet, the biosensor technology is reviewed focusing on its potential application to pathogen detection in the human food chain.

LIST OF ABBREVIATIONS

3D	Three-dimensional (referred to structure)
λ	Lambda; lysogenic phage that specifically infects *E. coli*
μg	Microgram; unit of mass equal to 10^{-6} grams
μL	Microlitre; unit of volume equivalent to 1^{-6} litre
μm	Micrometer; metric unit of length equal to 10^{-6} meters
Ab	Antibody
AP	Alkaline Phosphatase; one of the most widely used reporter enzymes
ATP	Adenosine 5'-TriPhosphate
BOD	Biochemical Oxygen Demand
BOT	*Clostridium botulinum* Toxin
BSA	Bovine Serum Albumin
CE	Capillary Electrophoresis
CFU	Colony Forming Unit; applies to viable/culturable bacteria
CT	Cholerae Toxin
CV	Cyclic Voltammetry
Da	Dalton; molecular weight unit
DNA	DeoxyriboNucleic Acid
DPV	Differential Pulse Voltammetry
dsDNA	Double Stranded DNA, in opposition to ssDNA
EDC	1-Ethyl-3-[3-Dimethylaminopropyl]Carbodiimide hydrochloride
ELISA	Enzyme-Linked ImmunoSorbent Assay
EU	European Union
FET	Field Effect Transistor
fg	Femtogram; unit of mass equal to 10^{-15} grams

FIA	Flow Injection Assay
FISH	Fluorescent *In Situ* Hybridization
FITC	Fluorescein IsoThioCyanate, one of the most used fluorophores
fmol	Femtomol; unit equal to 10^{-15} mol
FRET	Förster / Fluorescence Resonance Energy Transfer
g	Gram; unit of mass
HRP	HorseRadish Peroxidase; one of the most widely used reporter enzymes
HSV	Herpes Simplex Virus
EIS	Electrochemical Impedance Spectroscopy
IgG	Immunoglobulin G; the antibody type most widely used in immunodetection
IgE	Immunoglobulin E
IMS	ImmunoMagnetic Separation
IPTG	IsoPropyl-β-D-Thio-Galactoside; inducer of β-galactosidase production
ISFET	Ion-Selective Field-Effect Transistor
Ka	Affinity constant
Kd	Dissociation constant
L	Litre; volume unit
LB	Luria-Bertani Broth; the most widely used rich medium for bacteria culture
LOD	Limit Of Detection
LPS	LipoPolySaccharide
LRSP	Long-Range Surface Plasmon
M	Molar; concentration unit equivalent to 1 mole per litre
MAb	Monoclonal Antibody
MB	Molecular Beacon
Min	Minutes
MIP	Molecularly Imprinted Polymers
mL	Millilitre
MP	Magnetic Particles
mRNA	Messenger RNA
ng	Nanogram; unit of mass equal to 10^{-9} grams
NHS	N-HydroxySuccinimide
NIS	Non-faradic Impedance Spectroscopy
nm	Nanometer; metric unit of length equal to 10^{-9} meter
PAP	p-AminoPhenol

PAPG	4-AminoPhenyl β-D-Galactopyranoside
PBS	Phosphate Buffer Saline solution
PCR	Polymerase Chain Reaction
PFU	Plaque Forming Unit
pI	Iso-electric Point
pM	Picomolar; unit of concentration equal to 10^{-12} M
PNA	Peptide Nucleic Acid
QCM	Quartz Crystal Microbalance
TMB	TetraMethylBenzidine
RIA	RadioImmunoAssay
RNA	RiboNucleic Acid
rRNA	Ribosomal RNA
RT-PCR	Real Time PCR
SAM	Self Assembled Monolayer
SATA	N-Succinimidyl-S-AcetylThioAcetate
SAW	Surface Acoustic Wave sensor
SELEX	Systematic Evolution of Ligands by EXponential enrichment; the standard procedure for aptamer production /selection
SMCC	Sulfosuccinimidyl 4-[N-Maleimidomethyl]Cyclohexane-1-Carboxylate
SPE	Screen Printed Electrode
SPR	Surface Plasmon Resonance
ssDNA	Single Stranded DNA, in opposition to dsDNA
SWCN	Single-Walled Carbon Nanotubes
T4	Lytic phage that specifically infects *E. coli*
USA	United States of America
V	Volt; unit of electromotive force commonly called *voltage*

Chapter 1

PATHOGENS AT THE HUMAN ENVIRONMENT

ABSTRACT

Microorganisms have been an integral part of life on Earth since immemorial times. Since ancient times, humans learned how to take advantage of some bacteria and yeast for the production of a wide range of products. But history also showed humans the devastating effects produced by pathogen outbreaks. Overpopulation, denutrition, and insalubrity, among others, favored the propagation of plagues that have caused historically innumerable deaths. Nowadays, in spite of the medical advances, pathogens are still the main cause of deaths in the world. The ability of microorganisms to evolve extraordinarily fast, the advances in genetics engineering, the increasing mobility of the population worldwide, the unrestrained use of antibiotics, and the use of modified microorganisms *in field* are just setting the basis for new hazards. The iceberg top might be the sprouting of new bacterial strains showing multiresistance to the most widely used antibiotics in western hospitals, or the growing number of individuals affected worldwide by pathogens classically restricted to local environments.

In this book, we will focus on those pathogens potentially present at the human environment, which can cause outbreaks due to consuming contaminated water or food, and the different existing and developing strategies for their detection.

1.1. WATERBORNE PATHOGENS

Water is an essential resource for life which constant supply is put at risk with more frequency due to the continuously increasing population worldwide. Water is used by humans for many applications, including direct human use (drinking and cooking water), but also agricultural, recreational, household and industrial usage. Because of this, different regulations have been adopted for assuring water quality in the diverse contexts. Contamination of water by pathogens is a society's major health concern since a high number of people can be affected in a single outbreak. Furthermore, waterborne diseases are estimated to cause around 1.8 million deaths each year worldwide [1]. Although more prevalent, these epidemics are not unique to developing countries, where about 1.1 billion people have no access to proper drinking water [1]. In fact, the most important waterborne outbreak occurred to date was a Cryptosporidiosis outburst occurred in Milwaukee (USA) in 1993 [2] with more than 400,000 affected individuals. Ingestion of contaminated drinking water, followed by accidental ingestion of recreational waters, are the major transmission routes of waterborne pathogens, although consumption of contaminated agricultural products irrigated with infectected water has also been reported.

The pathogens most frequently isolated as the causative agents of waterborne outbreaks are *Cryptosporidium* and *Giardia* among protozoa [2-10]; *Campylobacter*, pathogenic *E. coli* (especially *E. coli* O157:H7), *Salmonella*, *Shigella*, and *Yersinia* as major representants of bacteria [4, 11-23]; and norovirus, rotavirus, and hepatitis A and E viruses [24-28]. A list of waterborne associated microorganisms, as well as the diseases caused by them, is presented in Table 1. However, many other species have been involved or are potentially waterborne transmitted. Not forgetting that a number of outbreaks are produced by unknown, or at least unidentified, ethiological agents. This can be often explained by the presence of viable but non-culturable bacteria and justifies the need for new analytical methods able to overcome the limitations of the currently available detection strategies.

The effective treatment of water, the protection of water resources and the establishment of periodic surveillance programs to assure the absence of pathogens in the water systems are needed to prevent and reduce the number of waterborne outbreaks. However, the emergence of new pathogens associated to the re-use of wastewaters for different usages after wastewater treatment is a new concern. Also, waste disposal after treatment, like residual sludge, onto agricultural lands can act as a new route for transmission of waterborne pathogens.

Table 1. Diseases, agents and symptoms associated with water-and-foodborne disease. Adapted from [29]

Disease	Agent	Symptoms
Acute diarrhoea	*Campylobacter jejuni*	Fever, diarrhoea, bloody stools
	Salmonella (non-typhoid)	Mild gastroenteritis, acute diarrhoea, fatal septicaemia (blood poisoning)
	E. coli	Fever, diarrhoea, bloody stools, uraemic syndrome
	Shigella	Fever, diarrhoea, bloody stools
	Vibrio vulnificus	Vomiting, diarrhoea, and abdominal pain
Yersiniosis	*Yersinia enterocolitica*	Fever, gastroenteritis with diarrhoea, abdominal discomfort
Listeriosis	*Listeria monocytogenes*	meningitis, meningoencefalitis, endocarditis, pneumonia
Brucellosis	*Brucella*	chills, headache, low back pain, joint pain, malaise, occasionally diarrohea
Botulism	*Clostridium botulinum*	Double vision, blurred vision, drooping eyelids, slurred speech, difficulty swallowing, dry mouth, and muscle weakness
Gastroenteritis	*Clostridium perfringens*	Diarrhoea, abdominal cramps, and nausea
Typhoid fever	*Salmonella typhi*	Fever, headache, appetite loss, nausea, diarrhoea, vomiting, abdominal rash
Cholera	*Vibrio cholerae*	Watery diarrhoea, vomiting, occasional muscle cramps
Legionnaire's Disease, Pontiac Fever	*Legionella pneumophila*	Malaise, headache, fever, muscle aches, pains, chills, cough, pulmonary symptoms
Viral Hepatitis	Hepatitis A and E viruses	Fever, chills, anorexia, abdominal discomfort, jaundice, hepatitis, headache
Gastroenteritis	Norovirus, Rotavirus, Adenovirus, Picornavirus, Astrovirus	Diarrhoea, discomfort, vomiting, malaise, headache, fever, muscle aches, pains, chills, cough, pulmonary symptoms
Staphylococcal food poisoning	*Staphylococcus aureus*	Nausea, vomiting, stomach cramps, and diarrohea
Cryptosporidiosis	*Cryptosporidium parvum*	Diarrhoea, abdominal discomfort
Giardiasis	*Giardia lamblia*	Diarrhoea, abdominal discomfort
Cyclosporiasis	*Cyclospora*	Loss of appetite, weight loss, stomach cramps/pain, bloating, increased gas, nausea, and fatigue
Toxoplasmosis	*Toxoplasma gondii*	Flu-like symptoms, swollen lymph glands, or muscle aches and pains, congenital defects (brain and eye) if mother infected
Amaebiasis	*Entamoeba histolytica*	Diarrhoea, abdominal discomfort

1.2. FOODBORNE PATHOGENS

The spectrum of foodborne outbreaks has substantially changed over time, mainly reflecting the changes in the cultural habits, as well as the technological and medical advances that limit or prevent contamination from occurring. For example, measures such as milk sanitation and pasteurization, regulation of shellfish beds, disease control in animals, or the interdiction to feed animals with uncooked garbage contributed to eradicate diseases like typhoid fever, tuberculosis, brucellosis, or trichinosis from industrialised countries, where they were common in the past [30]. Nevertheless, new microorganisms have gained prevalence as the causative agents of food poisoning. The spreading of ready-to-eat food or the extensive utilisation of antibiotics in veterinary, account among the reasons for this.

Foodborne outbreaks in humans usually happen as a result of eating vegetables and/or products of animal origin, such as raw eggs, poultry, pork, beef, fish, or their derivatives, which have been contaminated with a zoonotic pathogen. The suspected risk factors for vegetable contamination include the contamination of the irrigation wells with faeces from cattle and wildlife; direct exposure of the crops to wild animals and their faeces; and improperly composted animal manure used as fertilizer [31]. Still unclear is the relevance of pathogen internalization by the plants, either through the roots and plant vascular tissues or through vegetable/fruit surfaces into cracks and crevices. In the case of meat, contamination is often produced by manipulation during slaughtering, or over later meat manipulation and/or processing [32]. Specifically, the presence of foodborne pathogens in processed ready-to-eat products poses a serious threat to consumers, especially children, elders, and those individuals with compromised immune system. Most foodborne pathogens are microorganisms that are naturally shed in animal faeces, such as *Salmonella*, *Campylobacter*, *Listeria* or *E. coli*, and which under unhygienic conditions contaminate derived or neighbour food products. The range of foodborne pathogens, however, also includes a variety of viral pathogens and parasites, as well as marine bacteria able to produce biotoxins in fish and shellfish, and the self-inducing prions of the transmissible encephalopathies.

Each year, foodborne pathogens are responsible for large public health costs, including numerous sick days, medical expenses, and preventable deaths. For example, one in four persons is estimated to have a significant foodborne illness each year in the USA, where foodborne pathogens are the cause of 76 million of infections, 323,000 hospitalizations and 5,000 casualties per year [30], and at least 387,000 persons were affected by a zoonosis in the

European Union (EU) in 2005 [32]. Only in 2005, a total of 197,363 cases of infection by *Campylobacter* were reported by 22 EU member states, with total incidence of 51.6 per 100,000 inhabitants [32]. This makes campylobacteriosis the most frequently reported zoonotic disease in EU. In the USA, about 13 cases are diagnosed each year for each 100,000 inhabitants, with an estimate of approximately 124 annual casualties [33].

Salmonella alone infects over 40,000 individuals per year in the USA, where *Salmonella* serovars are associated with 26% of all foodborne diarrhoea that lead to hospitalization, causing approximately 400 annual deaths [34]. In EU, a total of 176,395 cases of human salmonellosis were reported in 2005, making it the second highest prevalence for a zoonosis: 38.2 cases per 100,000 inhabitants [35]. Significantly, salmonellosis can be transmitted by the consume of a wide variety of contaminated products, including homemade salad dressings and sauces, tiramisu, homemade ice cream, cookie dough, frostings, insufficiently cooked poultry and meat products, and raw or unpasteurized eggs, milk and other dairy products. As examples, an outbreak in early 2009 in the USA was originated by the consume of raw alfalfa sprouts contaminated with *Salmonella* serotype Saintpaul [36], and a later occurrence was attributed to a number of contaminated products, including instant non-fat dried milk, whey protein, fruit stabilizers, and gum-derived thickening agents [37].

Escherichia coli has a notorious reputation of causing food poisoning, mainly through contaminated poultry, vegetables and dairy products. The *E. coli* verotoxigenic strain O157:H7 is responsible for causing global disease outbreaks and potential death, with as few as 100 cells being sufficient to cause infection. For example, *E. coli* O157:H7 was identified as the causative agent of three different outbreaks in 2006 in the USA [38, 39]. Fresh spinach and lettuce were identified as the vehicles of illness, with 357 individuals affected, including 182 hospitalizations and 41 patients showing haemolytic uraemic syndrome, a type of kidney failure that can lead to death. More recently, on June 2009, contamination of pre-packaged Nestlé Toll House refrigerated cookie dough with *E. coli* O157:H7 were the cause of 69 infections in 29 different U.S. states, including 34 hospitalizations and 9 cases of haemolytic uraemic syndrome [40].

Listeria monocytogenes, which is naturally found in soil and water, is the causative agent of listeriosis. It is one of the most virulent foodborne pathogens, with 20% of the reported clinical infections resulting in death. Only in USA, *L. monocytogenes* is responsible for approximately 2,500 infections and at least 500 casualties annually [41]. Because vegetables can become contaminated from the soil or from manure used as fertilizer and animals can transmit the bacterium without being infected by it, *L. monocytogenes* can be found in a variety of raw foods, including uncooked meats, raw vegetables, unpasteurized milk and dairy products. In addition, and even when *Listeria* is efficiently killed by pasteurization and cooking, processed foods can become contaminated along the processing. For example, certain ready-to-eat foods, such as hot dogs and cold cuts, can suffer contamination between the cooking and packaging procedures, and soft cheese and cold cuts can be contaminated during manipulation at the deli counters.

Sensitive, specific and rapid detection of such pathogens is thus essential at production level to prevent their entrance into the human food chain. Detection of these new and re-emerging pathogens may be challenging because they can be sub-lethally injured due to the stress conditions suffered during treatment and storage, and may be missed using the traditional culture detection methods in spite of their ability to cause disease. To this end, the improvement of the existing microbiological techniques reported over the last decades, as well as the implementation of new analytical strategies, have lead to a more rapid, sensitive and specific detection of pathogens, which is important to diminish the public health risk. Some of these new techniques are discussed in chapters 3 and 4.

REFERENCES

[1] U.S. Centers for Disease Control and Prevention. Atlanta (2006). Safe Water System: A Low-Cost Technology for Safe Drinking Water. *World Water Forum, 4 Update.*

[2] W. R. Mac Kenzie, N. J. Hoxie, M. E. Proctor, M. S. Gradus, K. A. Blair, D. E. Peterson, J. J. Kazmierczak, D. G. Addiss, K. R. Fox, J. B. Rose and et al. (1994). A massive outbreak in Milwaukee of *cryptosporidium* infection transmitted through the public water supply. *New England Journal of Medicine, 331,* 161-167.

[3] R. G. Dantonio, R. E. Winn, J. P. Taylor, T. L. Gustafson, W. L. Current, M. M. Rhodes, G. W. Gary and R. A. Zajac (1985). A waterborne outbreak of Cryptosporidiosis in normal hosts. *Annals of Internal Medicine, 103,* 886-888.

[4] L. A. Duke, A. S. Breathnach, D. R. Jenkins, B. A. Harkis and A. W. Codd (1996). A mixed outbreak of *Cryptosporidium* and *Campylobacter*

infection associated with a private water supply. *Epidemiology and Infection*, *116*, 303-308.

[5] S. T. Goldstein, D. D. Juranek, O. Ravenholt, A. W. Hightower, D. G. Martin, J. L. Mesnik, S. D. Griffiths, A. J. Bryant, R. R. Reich and B. L. Herwaldt (1996). Cryptosporidiosis: An outbreak associated with drinking water despite state-of-the-art water treatment. *Annals of Internal Medicine*, *124*, 459-468.

[6] C. E. Lopez, A. C. Dykes, D. D. Juranek, S. P. Sinclair, J. M. Conn, R. W. Christie, E. C. Lippy, M. G. Schultz and M. H. Mires (1980). Waterborne giardiasis - a community-wide outbreak of disease and a high-rate of asymptomatic indection. *American Journal of Epidemiology*, *112*, 495-507.

[7] M. M. Marshall, D. Naumovitz, Y. Ortega and C. R. Sterling (1997). Waterborne protozoan pathogens. *Clinical Microbiology Reviews*, *10*, 67-85.

[8] P. D. Roach, M. E. Olson, G. Whitley and P. M. Wallis (1993). Waterborne Giardia cysts and *Cryptosporidium* oocysts in the Yukon, Canada. *Applied and Environmental Microbiology*, *59*, 67-73.

[9] T. R. Slifko, H. V. Smith and J. B. Rose (2000). Emerging parasite zoonoses associated with water and food. *International Journal for Parasitology*, *30*, 1379-1393.

[10] P. F. M. Teunis, G. J. Medema, L. Kruidenier and A. H. Havelaar (1997). Assessment of the risk of infection by *Cryptosporidium* or *Giardia* in drinking water from a surface water source. *Water Research*, *31*, 1333-1346.

[11] V. J. Dev, M. Main and I. Gould (1991). Waterborne outbreak of *Escherichia-coli* O157. *Lancet*, *337*, 1412-1412.

[12] K. V. Eden, M. L. Rosenberg, M. Stoopler, B. T. Wood, A. K. Highsmith, P. Skaliy, J. G. Wells and J. C. Feeley (1977). Waterborne gastrointestinal illness at a ski resort - isolation of *Yersinia enterocolitica* from drinking-water. *Public Health Reports*, *92*, 245-250.

[13] W. E. Keene, J. M. McAnulty, F. C. Hoesly, L. P. Williams, K. Hedberg, G. L. Oxman, T. J. Barrett, M. A. Pfaller and D. W. Fleming (1994). A swimming-associated outbreak of hemorrhagic colitis caused by *Escherichia-coli* O157-H7 and *Shigella-sonnei*. *New England Journal of Medicine*, *331*, 579-584.

[14] M. L. Rosenberg, J. P. Koplan, I. K. Wachsmuth, J. G. Wells, E. J. Gangarosa, R. L. Guerrant and D. A. Sack (1977). Epidemic diarrhea at

Crater Lake form enterotoxigenic *Escherichia-coli* - Large waterborne outbreak. *Annals of Internal Medicine*, *86*, 714-718.

[15] D. L. Swerdlow, B. A. Woodruff, R. C. Brady, P. M. Griffin, S. Tippen, H. D. Donnell, E. Geldreich, B. J. Payne, A. Meyer, J. G. Wells, K. D. Greene, M. Bright, N. H. Bean and P. A. Blake (1992). A waterborne outbreak in Missouri of *Escherichia coli* O157-H7 associated with bloody diarrhea and death. *Annals of Internal Medicine*, *117*, 812-819.

[16] F. J. Angulo, S. Tippen, D. J. Sharp, B. J. Payne, C. Collier, J. E. Hill, T. J. Barrett, R. M. Clark, E. E. Geldreich, H. D. Donnell and D. L. Swerdlow (1997). A community waterborne outbreak of salmonellosis and the effectiveness of a boil water order. *American Journal of Public Health*, *87*, 580-584.

[17] R. M. Clark, E. E. Geldreich, K. R. Fox, E. W. Rice, C. H. Johnson, J. A. Goodrich, J. A. Barnick and F. Abdesaken (1996). Tracking a *Salmonella* serovar typhimurium outbreak in Gideon, Missouri: Role of contaminant propagation modelling. *Journal of Water Supply Research and Technology-Aqua*, *45*, 171-183.

[18] G. F. Craun, R. L. Calderon and M. F. Craun (2005). Outbreaks associated with recreational water in the United States. *International Journal of Environmental Health Research*, *15*, 243-262.

[19] S. E. Hrudey, P. Payment, P. M. Huck, R. W. Gillham and E. J. Hrudey (2003). A fatal waterborne disease epidemic in Walkerton, Ontario: comparison with other waterborne outbreaks in the developed world. *Water Science and Technology*, *47*, 7-14.

[20] A. M. Maurer and D. Sturchler (2000). A waterborne outbreak of small round structured virus, *Campylobacter* and *Shigella* co-infections in La Neuveville, Switzerland, 1998. *Epidemiology and Infection*, *125*, 325-332.

[21] C. Arias, M. R. Sala, A. Dominguez, R. Bartolome, A. Benavente, P. Veciana, A. Pedrol, G. Hoyo and G. Outbreak Working (2006). Waterborne epidemic outbreak of *Shigella sonnei* gastroenteritis in Santa Maria de Palautordera, Catalonia, Spain. *Epidemiology and Infection*, *134*, 598-604.

[22] S. Martin, P. Penttinen, G. Hedin, M. Ljungstrom, G. Allestam, Y. Andersson and J. Giesecke (2006). A case-cohort study to investigate concomitant waterborne outbreaks of *Campylobacter* and gastroenteritis in Soderhamn, Sweden, 2002-3. *Journal of Water Health*, *4*, 417-424.

[23] J. M. Rangel, P. H. Sparling, C. Crowe, P. M. Griffin and D. L. Swerdlow (2005). Epidemiology of *Escherichia coli* O157 : H7

outbreaks, United States, 1982-2002. *Emerging Infectious Diseases*, *11*, 603-609.
[24] J. Hewitt, D. Bell, G. C. Simmons, M. Rivera-Aban, S. Wolf and G. E. Greening (2007). Gastroenteritis outbreak caused by waterborne norovirus at a New Zealand ski resort. *Applied and Environmental Microbiology*, *73*, 7853-7857.
[25] J. E. Kaplan, R. A. Goodman, L. B. Schonberger, E. C. Lippy and G. W. Gary (1982). Gastroenteritis due to Norwalk virus - an outbreak associated with a municipal water-system. *Journal of Infectious Diseases*, *146*, 190-197.
[26] M. Kukkula, P. Arstila, M. L. Klossner, L. Maunula, C. H. vonBonsdorff and P. Jaatinen (1997). Waterborne outbreak of viral gastroenteritis. *Scandinavian Journal of Infectious Diseases*, *29*, 415-418.
[27] A. L. Corwin, H. B. Khiem, E. T. Clayson, P. K. Sac, V. T. T. Nhung, V. T. Yen, C. T. T. Cuc, D. Vaughn, J. Merven, T. L. Richie, M. P. Putri, J. K. He, R. Graham, F. S. Wignall and K. C. Hyams (1996). A waterborne outbreak of hepatitis E virus transmission in southwestern Vietnam. *American Journal of Tropical Medicine and Hygiene*, *54*, 559-562.
[28] P. Coursaget, Y. Buisson, N. Enogat, R. Bercion, J. M. Baudet, P. Delmaire, D. Prigent and J. Desrame (1998). Outbreak of enterically-transmitted hepatitis due to hepatitis A and hepatitis E viruses. *Journal of Hepatology*, *28*, 745-750.
[29] D. L. Heymann (2001). Control of communicable diseases manual. 18th Ed. *American Public Health Association, Washington, DC 20001, USA*.
[30] R. V. Tauxe (2002). Emerging foodborne pathogens. *International Journal of Food Microbiology*, *78*, 31-41.
[31] M. P. Doyle and M. C. Erickson (2008). Summer meeting 2007 - the problems with fresh produce: an overview. *Journal of Applied Microbiology*, *105*, 317-330.
[32] B. Norrung and S. Buncic (2008). Microbial safety of meat in the European Union. *Meat Science*, *78*, 14-24.
[33] U.S. Centers for Disease Control and Prevention. Atlanta (2008), *Campylobacter*. http://www.cdc.gov/nczved/dfbmd/ disease_listing/ campylobacter_gi.html.
[34] U.S. Centers for Disease Control and Prevention. Atlanta (2008), Salmonellosis. http://www.cdc.gov/nczved/dfbmd/disease_listing/ salmonellosis_gi.html

[35] EFSA (2006). The community summary report on trends and sources of zoonoses, zoonotic agents and antimicrobial resistance and foodborne outbreaks in the European Union in 2005. *The EFSA Journal*, 94.

[36] United States Food and Drug Administration (2009). Raw Alfalfa Sprouts Linked to Salmonella Contamination. http://www.fda.gov/NewsEvents/Newsroom/ PressAnnouncements/ucm149570.htm.

[37] United States Food and Drug Administration (2009). Company Recalls Various Products Due to Potential Salmonella Contamination. http://www.fda.gov/NewsEvents/Newsroom/PressAnnouncements/ucm169471.htm.

[38] U.S. Food and Drug Administration (2006), Update: *E. coli* O157:H7 outbreak at Taco Bell restaurants likely over. FDA traceback investigation continues. http://www.fda.gov/bbs/topics/NEWS/ 2006/NEW01527.html.

[39] U.S. Food and Drug Administration (2007). FDA and states closer to identifying source of *E. coli* contamination associated with illnesses at Taco John's restaurants. http://www.fda.gov/bbs/topics/NEWS/2007/NEW01546.html.

[40] U.S. Food and Drug Administration (2009), FDA Confirms *E. coli* O157:H7 in Prepackaged Nestlé Toll House Refrigerated Cookie Dough. http://www.fda.gov/NewsEvents/Newsroom/PressAnnouncements/ucm169733.htm.

[41] U.S. Centers for Disease Control and Prevention, Atlanta (2008) *Listeriosis,*
http://www.cdc.gov/nczved/dfbmd/disease_listing/listeriosis_gi.html.

Chapter 2

CONVENTIONAL AND RAPID METHODS FOR PATHOGEN DETECTION

ABSTRACT

Traditionally, water and food industries have relied on the detection of bacterial indicator microorganisms for the prevention of faecal contamination, which denotes potential presence of pathogens in the system. Tests for detection of these pathogens have been conventionally based on classical culture methods, and the established indicator microorganism of faecal contamination has been *E. coli*. The use of these tests over the past century for the surveillance of water systems and the food chain has reduced efficiently the transmission of water- and foodborne pathogens to the population producing a marked decrease in those outbreaks over time. However, the technological advances produced in the last decades have prompted the development of new rapid methods that allow detection of target pathogens themselves with relative easiness and significantly shorter assay times [1]. Although the recommendations of the World Health Organization [2] emphasize that water monitoring does not necessarily equate pathogen monitoring, detection of pathogens can provide useful information for the development of monitoring models.

2.1. CULTURE-BASED METHODS

Detection of viable bacteria has been traditionally performed by measuring growth of individual microorganisms at more or less selective growth media. Growing of isolated bacterial colonies gives information on both the titters and the population diversity, because each species generates colonies of characteristic size, shape and profile. Colony counting on semisolid agar media is usually preceded by pre-enrichment in liquid media, which allows amplification of injured but viable cells and low prevalent pathogens potentially present in real samples. With this aim, hundreds of liquid and semi-solid culture media have been developed over the last decades, and ready-to-use products are commercialised by a number of providers. The accurate selection of growth media and culture conditions can lead to isolation of the whole bacteria population, certain bacterial groups, or even specific species and strains ([3]). For example, *E. coli* serotypes belonging to different pathogenic groups such as enteroaggregative *E. coli*, enterotoxigenic *E. coli*, and verotoxigenic *E. coli* have been linked directly to waterborne and foodborne outbreaks. Direct detection of these pathogenic groups normally involve plating onto Mac Conkey or EMB eosin agar, followed by testing for lactose positive colonies with different agglutinating antisera. In combination with a number of additional methodologies, culture allows the acquisition of supplementary information about the detected bacteria. The analysis of virulence genes can be performed by PCR of different colony sweeps from the agar plates. Biological assays including cell culture on Hep-2 or HeLa cells are used to determine the adherence pattern and toxin production of these virulent groups. Also, many immunological ELISA and RIA tests have been described to detect these toxins.

Nevertheless, and in spite of the high sensitivity and selectivity attained, detection of bacterial pathogens based on culture is tedious and often requires more than two working days. For instance, *Campylobacter* detection has been routinely performed by concentration of water samples using a 0.22 µm pore filter, followed by incubation for 4 hours in Preston Broth (a non-selective media containing lysed blood, trimethoprim, rifampicin, polymysin B and amphotericin) supplemented with ferrous sulphate, sodium metabisulphite, and sodium pyruvate, and additional incubation onto Preston agar plates for 48 hours in a microaerofilic environment. Detection of *Legionella pneumophila* requires also a pre-concentration step and incubation on buffered charcoal-yeast extract (BCYG) [4] for at least 5 days at 37°C and confirmation with agglutination tests. In the case of *Salmonella*, which is frequently associated with poor-quality water, selective enrichment is needed for the isolation of this microorganism from environmental samples. Three selective enrichment media are used in diagnostic laboratories, namely Tetrathionate Broth, Selenite Broth and Rappaport-Vassiliadis Medium. After isolation in selective media, isolates can be tested with commercial identification systems or with the traditional Sugar Iron Agar, Urea Broth and Lysine Iron Agar. It has to be noted that, because of the low titters that may be present in real samples and the possibility that bacteria are sub-lethally injured, concentration of large volumes and pre-enrichment in a non-selective medium, such as Buffered Peptone Water, is needed. As a result, the confirmation of the presence of this microorganism can take up to 5 days.

Detection of waterborne viruses is of higher complexity due to their low concentration in the majority of water and food samples and the difficulty for their enrichment since they are obliged intracellular parasites. Therefore, samples of at least 100 L for drinking water and no less than 10 L for recreational water are commonly processed in order to concentrate any viruses potentially present into a small volume. Additionally, a host cell where the virus can multiply is needed for their subsequent enrichment and detection. After concentration, detection of viruses can be performed by monitoring the cytopathic effect caused by infection of the host cell and enumeration of the plaques formed in a plaque assay. The *monolayer plaque assay* is the most widely used method for the enumeration of enteroviruses. Its performance is based on the formation of a confluent cell monolayer which, following infection by the virus-containing sample, is grown between two layers of semi-solid agar medium. This prevents spread of the viruses over the culture and, when diluted appropriately, ensures physical isolation of single infective units. Successive cycles of virus infection, proliferation and lysis of the host cells generate the formation of circular transparent plaques on the otherwise translucid cell monolayer, which are clearly visible to the naked eye. Another variation of this technique is the *suspended cell plaque assay*, in which the cells are suspended in the agar. In this way, the time required for the formation of the monolayer is reduced, while the number of receptors available for the infection by the viruses is simultaneously increased. Virus infectivity may also be assayed in *liquid culture* by exploiting the most probable number (MPN) method. In this case, different dilutions of the (virus-containing) sample are assayed in parallel by inoculation of a bacterial liquid culture. Bacterial proliferation under the different conditions serves to estimate the number of viruses initially present in the sample.

Attempts to automatise some these classical techniques have been reported. For example, *àCOLyte SuperCount* and *PetriScan* ® are automated colony counting systems provided by Synbiosis (Frederick, MD) and Spiral Biotech (Norwood, MA). *TEMPO* is an automated most-probable-number (MPN) system from BioMérieux (Marcy l'Etoile, France). Cell and laser scanning cytometry has also been formatted into a rapid detection system.

ChemScan RDI, developed by Chemunex (Maisons, Alfort, France), is based on direct labelling of viable microorganisms trapped on a membrane, coupled with a laser scanning and counting system.

2.2. ENZYMATIC ASSAYS AND IMMUNOASSAYS

In spite of their performance, increasingly strict regulations in fields such as food safety, health, and environment control, together with the demand for faster, more sensitive and easier to use analytical tools, have disclosed the limitations of some of these classical culture-based techniques [5, 6]. While very selective and sensitive, bacteria culture and plate counting rely on long assay times and have to be carried out by trained recruits at appropriate facilities.

In the last decades, traditional culture methods have been substituted, or at least complemented, by enzyme-based tests. Incorporation of specific enzyme substrates into selective growth media might lead to biochemical identification of bacteria. Good examples are the assays founded on the hydrolysis of chromogenic and fluorogenic substrates such as 5-bromo-4-chloro-3-indolyl-β-D-galactopyranoside (X-gal) by β-galactosidase or β-glucuronidase enzymes for detection of total coliforms and *E. coli*, respectively. Also, hydrolysis of methylumbelliferyl beta-D-glucuronide by β-D-glucuronidase, produced by all coliforms except enterohaemorrhagic *E. coli* O157:H7, makes colonies fluoresce blue when exposed to ultraviolet light. Some of these methods have been validated by standard organisations such as ISO or CEN [7-9]. In addition, a number of commercially available biochemical kits exists for confirmation of presence of specific bacteria. Most of them rely on detection of differential fermentation of a battery of carbohydrates using pH indicators in the media and/or the utilization of specific amino acids or enzyme substrates. For example, the automated systems *VITEK* and *MicroLog* provided by BioMérieux and BiOLOG (Hayward CA) are based on identification of bacterial *fingerprinting* according to their metabolism on a number of substrates and carbon sources and/or their susceptibility to antimicrobial agents. Nevertheless, results are obtained after sample incubation for 16-24 hours.

One of the molecular methods most widely accepted and used is the enzyme-linked immunosorbent assay (ELISA). ELISA is based on the utilization of polystyrene microtitter plates with at least 96 individual wells. The surface of the wells is modified with antibodies specific towards the target pathogen of choice. Incubation of a small volume of sample (usually 50-200 µl) leads to pathogen capture. The subsequent washing steps ensure than non-specifically bound sample components are removed. Detection is then performed by incubation of a second antibody labelled with, for example, a reporter enzyme. ELISA has been successfully applied to detection of whole cells, both viable and non culturable, as well as cell components and cells lysates. The whole procedure is performed within 4-8 hours, depending on the assay format, and accurate selection of the Abs used provides high levels of assay specificity. Moreover, ELISA is automatable in its whole length, making the assay compatible with the simultaneous study of multiple samples. For example, *Triturus Analyser*, *EIAFoss*, and *VIDAS* are automated ELISA platforms developed and commercialized by Diagnostics Grifols (Barcelona, Spain), Foss Electronics (Hillerod, Denmark), and BioMérieux, respectively.

In spite of its advantages, ELISA exhibits limited sensitivity, with typical detection limits in the range of 10^5-10^6 CFU/mL, depending on the Ab used and the sample matrix analysed. The other main drawback of ELISA is the potential interference by real sample matrices. Immunomagnetic separation (IMS) circumvents these difficulties by incorporating the Ab on the surface of paramagnetic particles (MP). Incubation of the MP with the sample leads to specific capture of the target pathogen, followed by concentration using a magnetic field. Captured bacteria can then be resuspended in the desired solution at the appropriate concentration before performing detection. IMS has been successfully coupled to subsequent culture on agar plate, colorimetric and fluorescent sandwich detection, and genetic detection among others. For example, detection of most protozoa involves first their concentration and separation from the sample matrix, usually by gradient centrifugation or immunomagnetic separation, followed by staining with fluorescently labelled antibodies and observation under an epifluorescence microscope [10]. The use of immunomagnetic separation techniques for the selective isolation of microorganisms has been incorporated into recent standards.

2.3. NUCLEIC ACID-BASED DIAGNOSTICS

Nucleic acid-based diagnostics is based on the detection of the pathogen DNA/RNA. Accordingly, other pathogen components, such as toxins or protein components, can not be studied. The four major techniques employed are enzymatic DNA restriction, fluorescent *in situ* hybridization (FISH), polymerase chain reaction (PCR), and fluorescence-based DNA microarrays

[11, 12]. Among them, PCR has generated significant levels of detection specificity and extraordinarily low detection limits. The main limitation of these methods is that their performance depends on the successful extraction of the pathogen genetic material and implies an important degree of sample pre-treatment and manipulation. A number of factors, such as the quality of the extracted DNA/RNA, the presence in the sample of inhibitors of the enzymes used for detection, or the co-extraction of nucleases, can significantly affect the final result and assay reproducibility. In addition, detection of the pathogen nucleic acids usually depends on the knowledge of its sequence in order to produce complementary probes for detection. Finally, these are destructive techniques and following analysis samples can not be studied by other means.

FISH is based on the utilization of fluorescently labelled DNA probes, which have to be complementary to specific target sequences present *only* in the pathogen nucleic acid. The ribosomal 16S rRNA is one of the preferred molecular targets. Interestingly, the hybridisation can be directly performed on intact fixed cells and can be studied by fluorescent microscopy or confocal laser scanning microscopy. This makes FISH quite a fast and simple technique for identification of both culturable and non-culturable bacteria. On the other hand, the procedure has to be accurately optimised in order to prevent non-specific staining of non-target sample components, insufficient fixation of the bacteria, or ineffective penetration of the probe during hybridization. Additional drawbacks may be the presence of target bacteria with too low rRNA content as to generate a positive, interference due to endogenous fluorescence, and instability of the fluorescent probes over time caused by the exposure to light during manipulation, which might contribute to generate decreased or lost signal intensity.

In PCR, DNA is amplified using a thermocycler by an iterative procedure that consists of heat-denaturation of the DNA template into ssDNA, annealing of specifically designed oligonucleotide primers to their complementary target sequences within the template ssDNA, and primer extension by a thermostable DNA polymerase. As a result, PCR produces at each cycle two copies from each target sequence unit present and the number of sequence units grows exponentially along the experiment. The final product can be visualized by electrophoresis. In the case of real-time PCR (RT-PCR), a fluorescent label is also added to the reaction. This allows monitoring the amplification of up to hundreds of samples in real time by detecting fluorescence intensity. RT-PCR also guarantees minimal sample manipulation after detection and minimises the risk of crossed contamination. The whole procedure takes just few hours, when accurately optimized can be extremely specific and sensitive, and allows detection of both culturable and non-culturable cells. Interestingly, PCR has been occasionally performed directly from clinical specimens with minimal sample handling, and can be nearly completely automated. On the contrary, the presence of polymerase inhibitors in the sample, as well as contamination by nucleases, can decrease PCR efficiency and generate false negative results.

PCR has started to settle down as a promising alternative for rapid detection of microorganisms. PCR can provide detection of a single cell in just some few hours, can be formatted for multiplexed detection of a number of non-related microorganisms and, when properly optimised and validated, can afford extremely high specificity. The commercialisation of real-time PCR equipments and assay kits additionally grant exceptionally easy operation and lower possibility of contamination between samples. For example, DuPont (Wilmington, Del., USA) offers the BAX® System, a real-time PCR equipment that uses ready-to-use reaction tubes containing tableted reagents and only requiring sample addition. A pallet of 9 commercial kits provides versatile detection in a variety of real sample matrices within 24 hours. However, the results obtained with this methodology must be taken cautiously, since it detects also dead cells and may produce false positive results, for instance after water treatment. Furthermore, the presence of inhibitory substances in environmental matrices may also generate false negative results. Although these limitations can be partially solved by targeting the messenger RNA and the addition of internal controls, the methodology becomes more laborious and time-consuming, and we are still talking about non-portable costly equipment that has to be run by specialised staff.

DNA microarrays are probably the best example to illustrate the recent efforts made in order to produce high throughput detection assays. Although DNA microarrays are not yet fully developed for pathogen detection, the numbers of works reported in the field anticipate that they might be available in a near future. The development of DNA microarrays is possible thanks to the increasing amount of sequence information that is being released for most pathogens. A DNA microarray consists of a physical substrate where a high number of tiny dots have been printed, each of them consisting on a different DNA probe. Sample incubation allows then to determine simultaneously the presence of multiple targets in a very short assay time. The recent advances if microarray technology for pathogen detection will be summarized later in chapter 4.

REFERENCES

[1] D.V. Lim, J.M. Simpson, E.A. Kearns and M.F. Kramer (2005). Current and developing technologies for monitoring agents of bioterrorism and biowarfare. *Clinical Microbioly Reviews*, *18*, 583-607.

[2] WHO (2006). World Health Organization guidelines for drinking water quality. Volume 1. Recommendations. 3rd Ed. World Health Organization, Geneva, 515 pp.

[3] W. Harrigan (1998). Laboratory methods in food microbiology. 3rd Ed. *Academic Press, San Diego, California.*

[4] P. H. Edelstein (1981). Improved semiselective medium for isolation of *Legionella pneumophila* from contaminated clinical and environmental specimens. *Journal of Clinical Microbiology*, *14*, 298-303.

[5] K.S. Gracias and J.L. McKillip (2004). A review of conventional detection and enumeration methods for pathogenic bacteria in food. *Canadian Journal of Microbiology*, *50*, 883-890.

[6] K.G. Maciorowski, P.Herrera, F.T. Jones, S.D. Pillai and S.C. Ricke (2006). Cultural and Immunological Detection Methods for Salmonella spp. in Animal Feeds-A Review. *Veterinary Research Communications*, *30*, 127-137.

[7] K. Helrich (1990). Official Methods of Analysis of the Association of Official Analytical Chemists (Chapter 17). In, *Microbiological Methods, 15 ed.* (425–497).Arlington, VA, Association of Official Analytical Chemists.

[8] N. S. Hobson, I. Tothill and A.P.F. Turner (1996). Microbial Detection. *Biosensors and Bioelectronics*, *11*, 455-477.

[9] C.W. Kaspar and C. Tartera (1990). Methods for detecting microbial pathogens in food and water. *Methods in Microbiology*, *22*, 497-530.

[10] Anonymous (1991). Drinking water: National primary drinking water regulations; total coliform proposed rule. *Federal Register 54* 27544-27568.

[11] K.B. Barken, J.A.J. Haagensen and T. Tolker-Nielsen (2007). Advances in nucleic acid-based diagnostics of bacterial infections. *Clinical Chimica Acta*, *384*, 1-11.

[12] E.A. Mothershed and A. M. Whitney (2006). Nucleic acid-based methods for the detection of bacterial pathogens: present and future considerations for the clinical laboratory. *Clinical Chimica Acta*, *363*, 206-220.

Chapter 3

BIOSENSORS: AN ALTERNATIVE TO TRADITIONAL METHODS

ABSTRACT

According to the convention, biosensors are "analytical devices incorporating a biological, biologically derived or biomimic material intimately associated with or integrated within a physicochemical transducer or transducing microsystem, which may be optical, electrochemical, thermometric, piezoelectric, magnetic or micromechanical" (http://www.biosensors-congress.elsevier.com/about.htm). Hence, biosensors are primarily composed of a sensing or transducing surface, which has been modified by incorporation of a biorecognition element specific towards the target under study. Subsequent capture of the target microorganism induces changes in the physical properties of the sensor surface and transduction efficiency across it, which is converted by the transducer into a measurable signal. Biosensors are consequently classified according to their signal transduction method (optical, electrochemical, calorimetric, piezoelectric) and/or according to the biological recognition element exhibited (immunosensors, DNA sensors, aptasensors).

Biosensors have been repeatedly proposed as a promising technology for the fast detection of pathogen microorganisms [1-6]. The fact that most biosensing strategies allow direct recognition of the target capture event in truly reagent-less assay formats, places biosensors ahead from sandwich-based strategies and/or methods making use of reporter or label components. Accordingly, direct transduction can generate results in extremely short assay times and allow monitoring in real time. Nevertheless, sensing based on a single capture event or biorecognition component might also generate poor specificity. In this respect, biosensor performance strongly depends on the optimisation of appropriate surface functionalisation and blocking protocols. As it will be described later in the text, a variety of strategies have been reported for the successful incorporation of biocomponents onto the materials predominantly used for biosensor fabrication: gold, carbon derivatives, silicon derivatives and indium tin oxide.

Biosensor development and fabrication can also take advantage from the recent advances in microfabrication, miniaturization and nanotechnology. At least part of the transduction formats (specially the electrochemical ones) are compatible with the production of miniaturised, compact, portable, rapid and sensitive technology at relatively low costs. This sets the basis for the future production of inexpensive lab-on-chip devices, potentially integrating all the required functions for sample pre-treatment and target detection, while guaranteeing minimal sample manipulation by the user. The exploitation of such lab-on-chip devices will facilitate *in situ* assay performance, even by not specifically trained staff, and under a variety of working conditions. Some attractive applications are real-time *in-situ* environmental monitoring and *in-situ* diagnosis in developing countries or isolated settings [7-10]. In addition, the operation of small-size tools entails the utilization of minute volumes of reagents and samples, little disposal of potentially dangerous residues, minimal environmental and health exposure, and correlates with enhanced biomolecule kinetics and accelerated assay times.

Regardless of what has been described above, the only biosensors that have been successfully commercialised and are presently used target simple molecules, such as glucose or alcohol [11-13]. However, an increasing number of works report on bacteria biosensing. Most of them describe immunocapture of whole cells or bacterial components (pili, spores, enzymes), or hybridisation of the pathogen nucleic acids.

Over the following sections, we will describe the biorecognition components most widely applied to pathogen biosensing (antibodies and nucleic acid probes), as well as those more recently incorporated (lectins, aptamers, phages, peptides). We will then revise the various strategies for surface bioengineering. We will finish by summarising the basis of the most frequently used transduction strategies.

3.1. BIORECOGNITION ELEMENTS IN SENSOR BIOENGINEERING: THE CLUE FOR BIOSENSOR SPECIFICITY

The accurate selection of the most appropriate biorecognition element is one of the most crucial steps in biosensor development. The biorecognition element is responsible for the selectivity and specificity of the biosensor. Thus, in the lack of a truly specific biocomponent, the developed biosensor will not provide selective detection. For example, the incorporation of polyclonal antibodies (PAb) will generally provide lower levels of biosensor specificity that the utilization of highly specific monoclonal antibodies (MAb), as well as using a wide spectrum receptor will hardy induce capture of a single bacterial species. Similarly, the limit of detection generated by a biosensor is directly determined by the affinity of the immobilised ligand for the target of choice.

In any case, the researcher will have to make the choice between biorecognition elements that have been raised against whole cells, or those produced against cell lysates or isolated components. In the former case, the element recognises carbohydrate or lipoproteins exposed on the microorganism surface and the biosensor will provide better results for the direct detection of whole cells. In the latter, the ligand will recognise cell lysates, and thus cell components such as nucleic acids, enzymes, structural proteins, pili or spores, rather than whole cells. Accordingly, biosensor optimal performance might be attained only if disrupted cells are tested, more than probably requiring sample pre-treatment. Nonetheless, those few transducing strategies that display poor detection of whole cells (such as SPR, as it will be exposed later in the text) may provide improved results on lysate bacteria than in undisrupted cells.

Independently of the biorecognition element used, it has to be in intimate contact with the transducer element. Hence, it has to be incorporated onto the sensing surface and this modification should preserve as much as possible the biocomponent integrity and functionality. In addition, the bioengineered surface has to be inert and biocompatible, so that sample composition is not affected and remain constant over time, as to guarantee a stable signal baseline. Different immobilisation techniques have been successfully applied to anchor the biorecognition element to the transducer, including random physisorption, covalent immobilisation and avidin-biotin affinity capture onto the materials most widely used for biosensor fabrication, such as gold, carbon, and silica derivatives [3, 14-22]. The orientation, distribution and density of the receptors strongly determine the sensitivity of the biosensor and, when possible, have to be carefully optimised [23]. Besides, the most frequently used bioreceptors for detection of microorganisms are made of protein or nucleic acid, which are very sensitive to denaturation after binding onto a solid substrate (i.e. the transducer element). The long-term stability of the receptor and the receptor-modified surface, as well as the possibility to reuse them along time, are two additional parameters that should be also taken into account. The following section describes the most commonly used biorecognition elements, including antibodies and DNA probes, but also considers alternative biocomponents that are gaining a place in biosensor development, such as aptamers, peptides and phages.

3.1.1. Protein Based Biorecognition Elements

3.1.1.1. Antibodies

The use of antibodies (Ab) for immuno-detection of specific targets was first reported by Yallow and Berson in 1959 [24]. Since then, Ab have become the most widely used biorecognition elements in bioanalysis and a high number of biosensors for pathogen detection rely on immuno-capture. Ab have shown high sensitivity and specificity, not forgetting that Ab against an extensive number and variety of targets are commercially available. In addition, numerous protocols have been demonstrated successful for both their immobilisation onto surfaces and their chemical modification (for example with fluorescent, colorimetric and enzymatic labels). One of the main drawbacks of using Ab for biosensing is related to their relatively big molecular size (150 KDa). Thus, once immobilised onto a sensing surface, Ab can impair an important level of physical blocking that might interfere with signal transduction in some sensing formats. On the other hand, the fact that a number of biosensors are label-less and depend on a single immunocapture event (by opposition to sandwich assay formats), conditions sensor specificity to the availability of Ab truly specific for the target under study. Finally, Ab are extremely difficult to regenerate. This hampers the production of re-usable immunosensors in favour of disposable devices, contributing to increase the final production cost. The structure of IgG, the most widely used Ab type, is illustrated in Figure 1.

Polyclonal antibodies (PAb) are the most widely used Ab in biosensing. This is mainly due to the facts that PAb are cheaper than MAb and survive in better shape both chemical modification and immobilisation onto surfaces. PAb are produced by inoculation of the target of choice into a mammal. The "intruder" triggers the animal immune response, leading to the production of a number of Ab species. This means that each preparation of PAb contains a mixture of Abs recognising different epitopes present on the target. Production is consequently highly variable due to batch-to-batch differences. Additionally, PAb are hard to produce against small molecules, which may not induce an appropriate immunogenic response, and towards compounds that are toxic for the animal. PAb are useful when not a too high specificity is searched, such as simultaneous detection of various bacterial serotypes.

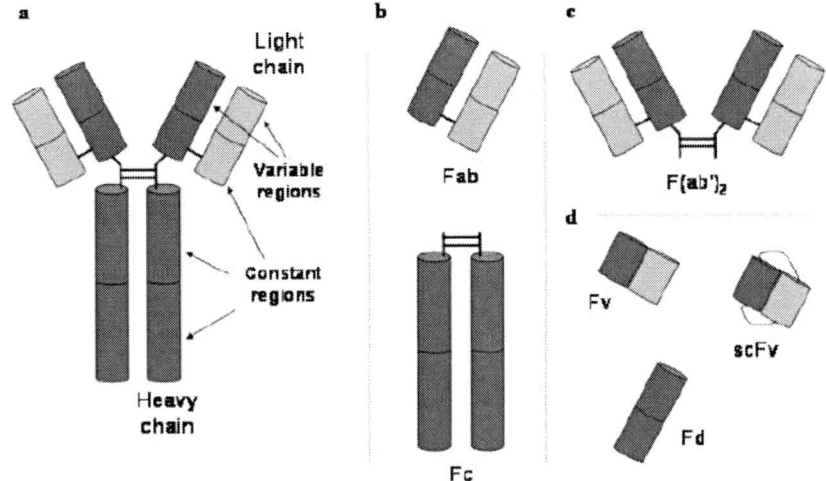

Figure 1. Structure of Ab and Ab fragments of different types. (a) Each IgG Ab is composed of two identical heavy chains and two identical light chains, which are attached by disulphide bonds. Each chain exhibits a highly variable domain close to the amino terminal (where are located the target-binding domains), and constant regions at the carboxyl terminal extreme. (b) Ab enzymatic digestion using *papain* generates two Fab fragments, each containing a whole light chain and half a heavy chain, and an Fc fragment. (c) *Pepsin* digestion generates a F(ab')2 fragment. (d) Among other known types of fragments, Fv is composed of the two variable regions from a heavy and a light chain and, even if it is able to bind to an antigen, it is unstable. A single-chain Fv fragment (scFv) is a stable variant of Fv commonly produced by recombinant technology, in which a polypeptide linker connects the two variable regions. An Fd fragment contains the amine terminal half of the heavy chain.

Monoclonal Ab (MAb), on the other hand, are produced *in vitro* by hybridomas, a methodology reported in the mid 70s [25]. The establishment of a hybridoma consists in fusing an immortal cell line with a single blood cell precursor, which is isolated from the plasma of an animal that has been immunised with the target of choice. In this way, each hybridoma produces a single type of Ab able to bind a single epitope on the target. MAb are thus characterised by their high specificity. The procedure allows in this case production of relatively high amounts of MAb over time, with little batch-to-batch differences, and reproducible specificity and sensitivity. Nevertheless, MAb suffer higher levels of inactivation induced by chemical modification and immobilisation onto physical substrates than their counterpart PAb. In addition, their highly appreciated selectivity is often incompatible with detection of variable microorganisms and MAb cocktails should be used instead.

A more recent alternative consists in the production of Ab fragments by Ab partial digestion. The advances in recombinant technology have even enabled *in vitro* production and selection of some of these variants, obtained by cloning of the sequences codifying for the Ab binding sites into expressing hosts (transformed *Escherichia coli*, phage displayed libraries). The various products that have been described - Fab, Fab', F(ab')2, Fv, and sc-Fv, depending on their size and components- are depicted in Figure 1. Ab fragments not only keep target binding ability, but are much smaller in size than native Ab [26, 27]. The incorporation of such small-size Ab variants to sensor development is expected to favour the immobilisation of a higher number of functional molecules, generating better surface coverage and improved detection sensitivity, to induce lower levels of surface blocking, and to favour device miniaturisation [28]. Nevertheless, only a small number of reports have described their use to date [29]. This is presumably due to their low stability and their still limited commercial availability.

The easiest and fastest strategy for Ab immobilisation onto a sensor surface is via random deposition, an approach that has generated exceptionally good results for the detection of whole cells [20]. Nonetheless, most authors defend that Ab integrity and functionality is better preserved if immobilisation is directed. A variety of such approaches have been described, including chemical conjugation or crosslinking of the Ab to alkanethiol or silane monolayers (previously self-assembled onto gold or silica surfaces respectively), Ab affinity capture by protein A/G, and (strept)avidin binding of biotinylated Ab [15, 30-34].

3.1.1.2. Bacteriophages

Bacteriophages are viruses that infect bacteria with high specificity. They bind to determined receptors that are present on the cell wall or in the sexual pili of the bacteria. Because of the easiness of production and manipulation, compared to antibodies, they are becoming popular in the biosensing technology [35, 36]. In this respect, phages can be directly grown by just infecting an appropriate bacterial culture, and can be purified by (ultra)centrifugation and/or filtration (Figure 2). The whole procedure needs relatively simple facilities, minimal manipulation skills, and does not require the use of animals.

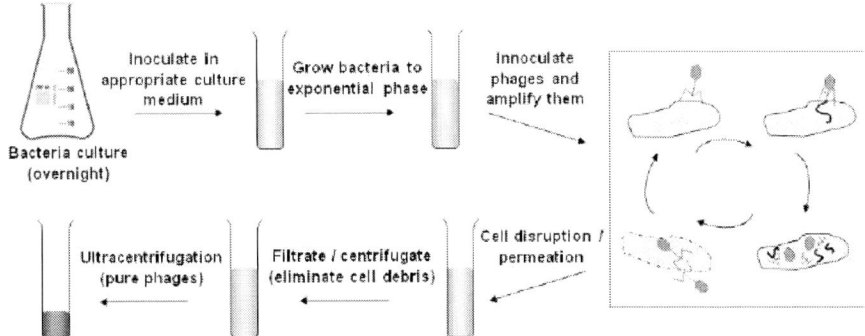

Figure 2. Example of bacteriophage production procedure. Phage culture / amplification can be carried out at the laboratory by simple infection of the host bacteria in a solution or culture media containing all the ions / components required for phage infection. The infection cycle (inset) starts with phage binding to specific receptors on the surface of the host bacteria, followed by injection of the phage genome. Bacterial enzymes will then produce copies of all the phage components (genome and structural/enzymatic proteins). The cycle finishes with the assembly and liberation of new phage units into the medium, often by cell disruption / lysis.

Modification of sensors by random deposition of bacteriophages has generated excellent results, but directed conjugation on SAM-modified surfaces has been also reported. Some few works have even described the production of genetically modified phages which naturally incorporate biotin on their surface allowing directed capture by biotin-binding proteins [37, 38]. Bacteria capture involves in this case physical interaction between the phage and specific receptors on the cell surface, followed by injection of the phage genome into the bacteria. Accordingly, once "used", phage-modified surfaces shouldn't be regenerated and/or re-used.

The advances in phage displayed peptide technology have triggered new applications of engineered bacteriophages for biosensing purposes. Phage display, described more than two decades ago [39], consists in the genetic engineering of bacteriophages to express different proteins/peptides on their capsid surface. With this method, a high number of bacteriophages can be screened simultaneously by exposing them to a target bacterium. After the screening, the bacteriophages showing the highest affinity constant (Ka) can be selected and produced in large quantities. Although easy to produce, phages have the shortcoming of exhibiting in average a lower Ka than their antibody counterparts. On the other hand, phage displayed peptides can be produced against ligands potentially harmful for an animal that can not be inoculated to generate antibodies [40]. Because of the high potential of this technology in biosensing, a lot of research is being devoted to the field [41].

3.1.1.3. Lectins

Carbohydrates have also been targeted for the specific detection of pathogenic bacteria [42]. Gram negative bacteria have a unique profile of glycoproteins in the lipopolysaccharide (LPS) and/or the capsule, which are serotype specific (O and K antigens, respectively) [43]. Detection of such glycoproteins can be accomplished by the use of lectins. Lectins are a special type of proteins which are ubiquitously produced in nature by animals, plants and microorganisms, and possess a highly specific carbohydrate binding domain. Illustrative examples are the lectins used by some viruses to attach themselves to the cells of the host organism during infection (such as hemagglutinin).

Characterisation of the glycoprotein profile using a battery of lectins can therefore serve to identify pathogenic bacteria. Lectins have been successfully used in microarray assay formats to detect the different glycan profiles present in the LPS of several pathogenic bacteria [44]. Some other few works also investigate the use of lectins for bacteria biosensing [45-48].

3.1.1.4. Antimicrobial Peptides

Antimicrobial peptides (AMPs) are defence peptides generated by the host immune response after being infected by a microorganism. Most of these peptides behave as potent, broad spectrum antibiotics. Thus, unlike other bioreceptors such as antibodies, they can recognise simultaneously a number of different bacteria. Because of that, several works reported on the utilisation of AMPs to characterise the binding pattern of various bacteria to them [49, 50]. The potential applicability of several different types of peptide-based bioreceptors to pathogen biosensing has been recently reviewed [51].

3.1.2. Nucleic Acid Biorecognition Elements

3.1.2.1. Nucleic Acid Probes

Nucleic acid probes are single-chain DNA or RNA fragments that are used to bind, by base pairing, complementary sequences present in the nucleic acids of the target microorganism. Detection specificity relies in the selection of target sequences that are known and are highly specific for the pathogen under study. Only in this way the design and synthesis of the specific complementary probes is possible. Interestingly, DNA/RNA probes are produced completely *in vitro*, either by chemical synthesis or by PCR. This facilitates the incorporation of functional groups/molecules during the synthesis procedure (for example, biotin or –SH groups to improve immobilisation, fluorophores or electrochemical labels to provide direct detection, or –NH$_2$ or –SH groups for subsequent conjugation). In spite of its sensitivity to enzymatic degradation, DNA is more resistant to manipulation and storing conditions than proteins. Additionally, oligonucleotide probes can be regenerated by thermal melting, allowing the production of truly re-usable sensing surfaces [52].

DNA sensors have not surpassed immuno-sensors in the biosensor market because of some handicaps. Among others, DNA sensing requires the pre-treatment of the samples as to extract DNA. This is, not only a critical step, but a procedure that liberates also a high amount of undesired components and results extremely difficult to integrate into microdevices. However, they have some characteristics that make them very promising biosensors [53]. For instance, the construction of microarrays (also known as DNA chips), which are miniaturised platforms allowing multiplex detection of different target genes, have great potential to construct biosensors to target different pathogens at the same time.

Nucleic acid probes can be attached easily to different transducer elements, compared to antibodies, and form relatively high stable biorecognition layers. The main techniques used for immobilisation of the nucleic acid probes onto the transducer element are similar to those used to anchor antibodies for the construction of immunosensors: adsorption, covalent-binding and avidin-biotin interactions [54]. Probe entrapment by electropolimerisation of a conducting polymer has also been reported [55, 56]. Additionally, oligonucleotides can be modified with SH$_2$ terminal groups and be directly self-assembled on gold surfaces.

Transduction of the hybridisation signal can be performed with electrochemical, optical or with mass sensitive sensors (see [54], for review). In this context, target amplification using PCR can provide, not only improvement of the detection limits, but also incorporation of signal amplifiers (electroactive molecules, biotin, metal or magnetic nanoparticles). This possibility has been successfully exploited for piezoelectric and electrochemical sensing, in some cases incorporating completely integrated microfluidics, and reporting detection limits down to 10^2-10^3 CFU/mL for *E. coli* [57, 58]. Electrode microarray technology has also been reported for amperometric and voltammetric multiplexed detection of up to 5 bacteria, in the first case with no need of previous amplification [59, 60]. Finally, the probe-target hybridisation event increases the interface negative charge, and thus negatively affects electron transfer across it. In this way, the hybridisation can be easily monitored using electrochemical biosensors [61].

3.1.2.2. Peptide Nucleic Acids

Peptide nucleic acids (PNA) are, as shown in Figure 3, artificially synthesised DNA mimics that lack the negatively charged sugar backbone of DNA [54, 62, 63]. They share the same characteristics that nucleic acid probes in terms of specificity and reusability, as described above, and also exhibit the ability to form PNA-DNA and PNA-RNA duplexes [64, 65]. Moreover, the PNA hybrids display higher thermal stability than native nucleic acids because of their neutral backbone. This characteristic has been exploited to develop PNA-based assays to detect single base pair mismatches. Detection is based in this case, in the fact that a single mismatch in DNA-PNA hybridisation is more unstable than in DNA-DNA hybridisation [66]. Moreover, PNA show stability to a wide range of temperatures and pH that DNA does not tolerate, and the addition of salts is not necessary to perform the hybridisation because the uncharged nature of the PNA molecule impairs no repulsion of the complementary strand [67]. For the reasons exposed above, PNA have been widely used and are promising bioreceptors for the fabrication of DNA sensors for pathogen detection in both electrochemical and optical transduction formats [67-69], and also for the development of microarray based sensors [70].

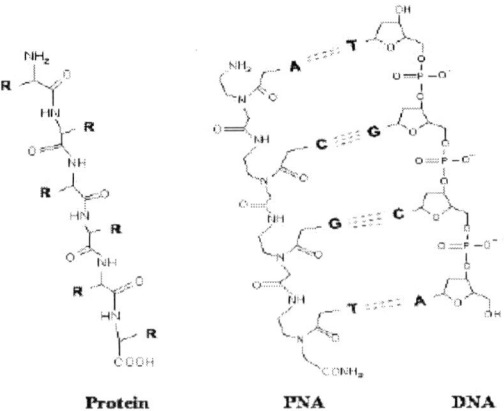

Figure 3. Chemical structure of peptide nucleic acids. Scheme showing the chemical structure and backbone characteristic of proteins, DNA and peptide nucleic acids (PNA). PNA and DNA of complementary sequence can hybridise by base pairing.

3.1.2.3. Aptamers

Aptamers were described for the first time in 1990 [71-73]. They are artificial nucleic acid ligands, produced by exponential enrichment of a combinatorial nucleic acid library against the target of choice. The procedure, coined SELEX (systematic evolution of ligands by exponential enrichment), consists in an *in vitro* iterative process that starts with the incubation of the nucleic acid library with the target. If nucleic acid sequences with the ability to bind to the target exist, they are recovered and amplified in the successive binding, recovery and re-amplification steps. Target recognition and binding usually depends on the aptamer folding into specific 3D structures (Figure 4). For this reason, aptamer immobilisation and/or labelling has to be carried out through long spacers in order to allow its proper folding. In the same way, optimisation of the binding buffer and binding conditions might have to be carried out for each aptamer in order to ensure optimal performance. Up to date, aptamers towards targets as varied as small molecules, peptides, proteins, and even whole cells, viruses and spores, have been reported. Some few works summarize the attempts performed to exploit aptamers for pathogen biosensing [74].

When compared to more classical approaches, such as Ab, aptamers are produced via a relatively simple procedure which is carried out completely *in vitro*. At least in theory, this property should make possible aptamer production against small, non-immunogenic or toxic targets. Apart of this, aptamers are characterised by smaller size and higher chemical simplicity than protein biocomponents, characteristics that favour regeneration, as well as immobilisation and labelling. The main drawback, on the other hand, is the high sensitivity of nucleic acids to nuclease attack. This problem has been solved via the production of *spiegelmers*, which are "mirror image aptamers" [75, 76], and via generation of chemically modified aptamers [77]. Both species have been demonstrated to be poorly or not hydrolyzed by nucleases, and survived for up to a few days in the *in vivo* studies.

Aptamer immobilisation can be carried out by direct self-assembly of thiolated species onto gold surfaces. However, directed anchoring by cross-linking onto self-assembled monolayers and streptavidin capture of biotinylated aptamers are often preferred, in order to produce surfaces more resistant against non-specific adsorption. Because efficient target capture often depends of the aptamer folding into the appropriate three-dimensional structure, aptamer immobilisation or modification should be always performed using long spacers or cross-linkers.

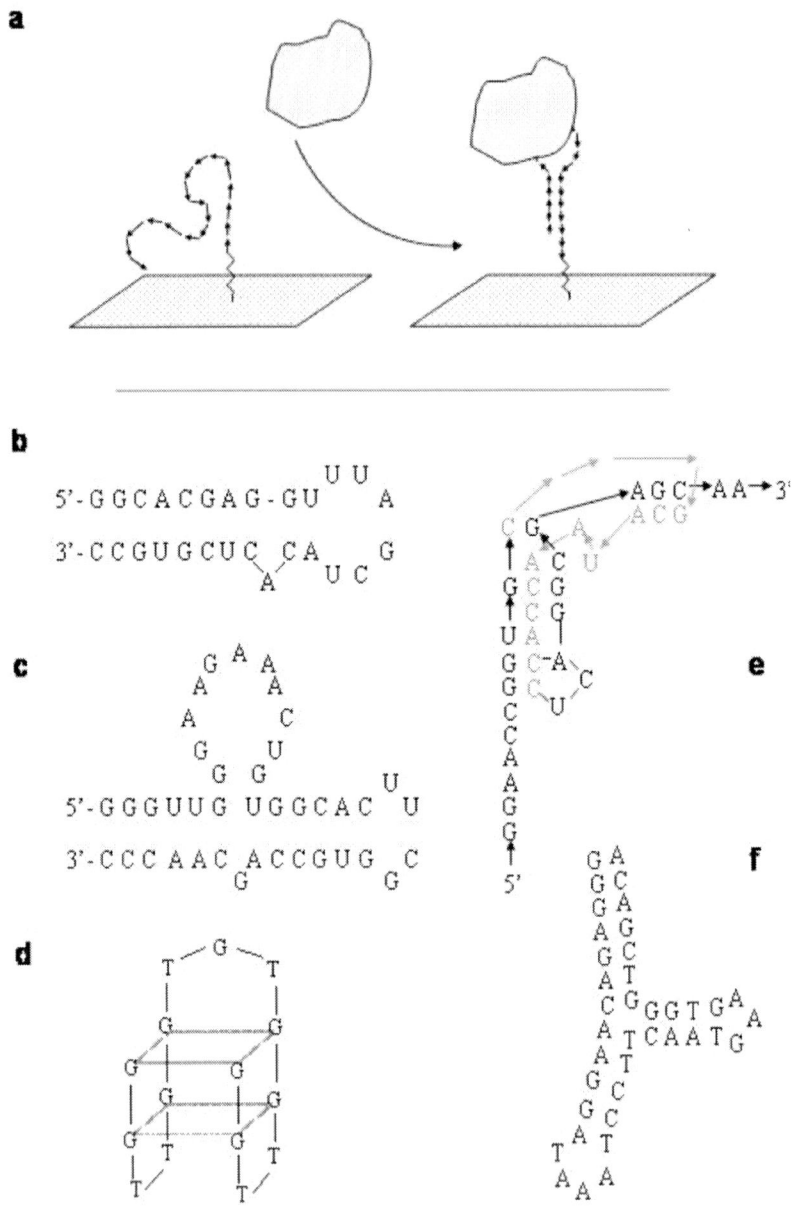

Figure 4. Aptamer folding. (a) Aptamers are nucleic acid probes that fold into a 3D structure which is responsible for (non-nucleic acid) target binding. Some of the most widely characterized 3D structures are the hairpin (b), stem-and-loop (c), quadruplex (d), pseudoknot (e), and the three-stem junction (f).

3.1.3. Whole Bacterial Cells as Biorecognition Elements

Immobilisation of bacterial cells to be used as sensing elements has been traditionally applied to the detection of toxicity and BOD (Biochemical Oxygen Demand) for wastewater treatment monitoring [78-80]. In recent years, the use of whole-cell biosensors has also been reported for the detection of bacteriophages as an indicator of bacterial presence. In the works reported, impedance and surface plasmon resonance biosensing is mostly based in the monitoring of cell integrity (for example, detection of cell disruption) and/or biofilm growth [81-83]. Alternatively, metabolic parameters can be studied electrochemically, such as production of metabolites or oxygen consumption by the living cells attached to the sensor surface before/after infection.

The main limitation of this type of sensors is that the bacterial film used as sensing element can not be regenerated after sample exposure, as happens in toxicity and BOD biosensors. For example, following phage infection/detection, bacteriophages might remain in a latent form in the surviving bacterial cells and induce later lysis in the bacterial film. It may also happen that some cells become resistant to these phages after long-term exposure to them. Therefore, the main shortcoming of these biosensors at the moment is that they are single use biosensors.

3.1.4. Biomimetic Receptors

Biomimetic receptors are artificially fabricated receptors that mimic a bioreceptor that occurs in nature. A good example are the molecular imprinted polymers (MIP), which are intended to mimic antibodies [84, 85]. This type of receptor is synthesized by polymerization of a determined polymer from its monomeric form in the presence of the target that is intended to bind (Figure 5). After polymerization occurs, the target can be removed by different extraction methods, leaving and empty cavity that will be later available for the recognition of new target molecules. The advantage of this approach relies on the fact that the newly synthesized receptor display several binding sites in one single receptor molecule, which are available for the analyte, allowing a magnification of the signal to be detected. However, the development of biosensors for microbial detection based on MIP is still challenging and only few reports have been documented [86, 87].

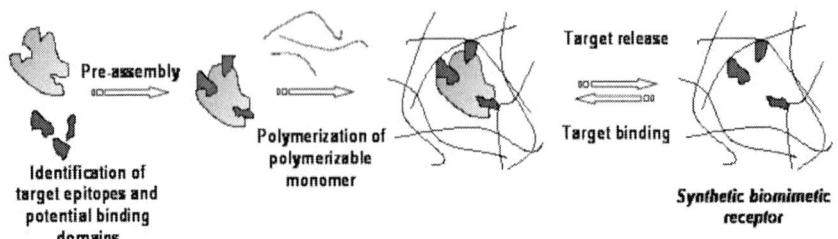

Figure 5. Example of production procedure of a biomimetic receptor. Biomimetic receptors are synthetic materials produced to specifically bind a certain ligand. Their production starts with the accurate characterisation of the target molecule, in the search for its epitopes, as well as the identification of natural binders (such as Ab, receptors, etc.) and their binding domains. Production involves polymerization of an appropriate monomer using the target units as a template and the binding domains for their internalization. Release of the template leaves an internal cavity with ability to subsequently bind target molecules.

3.2. FUNCTIONALISATION OF THE SENSING SURFACE: TOWARDS BIOSENSOR SPECIFICITY AND ROBUSTNESS

As it has been previously exposed, biosensor development strongly relies in the successful functionalisation of the electrode surface in order to integrate the selected biorecognition elements. The reason is that most of the parameters used to evaluate biosensor performance (named sensitivity, dynamic range, reproducibility and response time) depend on how far the original properties of the bioreceptor prevail after its immobilisation. Therefore, and independently of the strategy exploited for surface functionalisation, it has to provide biorecognition elements immobilised in a stable way, which retain both accessibility for the target molecule and its recognition ability. In addition, both the sensor surface and the biocomponent cover incorporated during functionalisation should be inert, biocompatible, and stable enough as to guarantee that sample composition/integrity will not be modified in any way and that a constant signal baseline will be generated. The following sections address the most common immobilisation techniques, including random physisorption, silanisation and SAM formation coupled to biomolecule cross-binding or covalent bonding, entrapment into polymers or membranes, and capture by affinity proteins.

3.2.1. Bioreceptor Random Physisorption

Random physisorption is, with no doubt, the easiest and fastest strategy for biomolecule immobilisation onto physical substrates. It consists on just depositing a small volume of a solution containing the bioreceptor of choice onto the surface that has to be modified, and leaving them to interact in a completely random way. When the biocomponent approaches the solid–liquid interface, where the medium conditions are different from the bulk solution, it interacts with the interface and rearranges in the search for a new "most stable structure", which is in most cases different from the most stable structure in solution. Depending on the characteristics of both the biocomponent and the surface, the initial interaction driving forces will be hydrophobic (for hydrophobic surfaces such as polystyrene) or electrostatic (for more polar substrates such as silica). However, because complex receptors such as proteins can be composed of series of hydrophilic/hydrophobic and charged/uncharged segments, biomolecule adsorption might be subsequently stabilised by a combination of hydrophobic interactions, hydrogen bonding and/or Van der Waals forces. Hence, the final product will be highly dependent on each individual protein-surface involved and will be highly stable [88].

Adsorption on gold surfaces of bioreceptors containing free -SH groups and/or S-S bonds, on the other hand, profit from the strong cross-binding between these functional groups and gold. For example, it has been demonstrated that Ab physisorption onto gold surfaces provides protein covers which are stable and functional for days [20]. Some authors defend that molecules such as Ab perform better if immobilisation is partly directed, for example by reduction of the inter-chain S-S bonds to generate free -SH groups, or by incorporation of long-chain spacers [19, 21].

In contrast, random physisorption may correlate with a certain degree of biocomponent partial denaturation, which could negatively affect its structure and/or function [89]. At least one report describes activity losses of up to 90% for physisorbed Abs, attributed to a combination of (i) loss of binding sites caused by denaturation, (ii) binding site hiding in poorly oriented molecules and (iii) steric hindrance as a result of molecule crowding and aggregate formation [90]. However, thanks to the high amount of molecules immobilised, the success rate is usually high enough to ensure reasonable assay performance. In the case of whole bacteria detection, physisorbed Ab have been described to perform even better than Ab immobilised via more complex strategies, such as cross-binding onto self-assembled monolayers

(SAM) [15, 20]. Accordingly, a significant number of biosensors reported for detection of different types of targets rely on biocomponent random physisorption thanks to its simplicity [91-97].

Independently of its potential advantages, the applicability of random physisorption is limited when certain biocomponents are to be immobilized. For example, physisorption may show poor performance for the immobilisation of small molecules (such as peptides or biotin), which may require previous conjugation to a bigger carrier molecule. In the same manner, random physisorption should be avoided for the immobilisation of a few bioreceptors, such as MAb which, due to unknown reasons, show worse performance following unspecific adsorption than PAb [90], and aptamers, whose performance depends on their folding into the appropriate three-dimensional structures.

3.2.2. Self Assembled Monolayers (SAM) and Bioreceptor Self Assembly onto Metal Surfaces

Self assembled monolayers (SAM) are formed by spontaneous organisation of thiolated molecules onto the surface of several metals, including gold, silver, copper, palladium, platinum and mercury (Figure 6) [98-101]. Although the chemical reaction involved is poorly understood, the product generated is highly stable. Furthermore, the fact that thiolated molecules of different length, composition and terminal groups can be exploited, allows the formation of organic surfaces whose composition, structure and properties can be varied rationally. This has lead several authors to propose SAMs as a versatile platform for sensor functionalisation [16, 102, 103]. In this respect, SAMs are easy to prepare and functionalize in any ordinary chemistry laboratory, can be formed on surfaces of any size and shape, and allow incorporation of all kinds of biocomponents via a number of well known chemical strategies (as it will be later revised in section 3.2.4).

Gold has become the standard substrate for biosensor functionalisation via SAM formation and modification for various reasons. Gold is easy to obtain, it is easy to pattern using photolithography, and it is bio-compatible. For SAM formation, the thiolated molecule is diluted in a suitable polar solvent (usually ethanol) to a concentration in the high micromolar to low millimolar range. The metal surface is then immersed into this solution and the reaction is allowed to proceed for a few minutes or up to several hours. Even if a few minutes appear to be sufficient for SAM formation, especially in the case of self-assembly of short chains, longer incubations generate higher molecular density and more ordered structures. This is particularly important when using long-chain alkanethiols, which may require self-assembly times of the order of hours. Apart of this, SAM formation is affected by a number of parameters, including the type of solvent used, the temperature at which the reaction takes place, the concentration, purity and structure of the thiolated molecule, the incubation time duration, and the cleanliness of the metal substrate [30, 104].

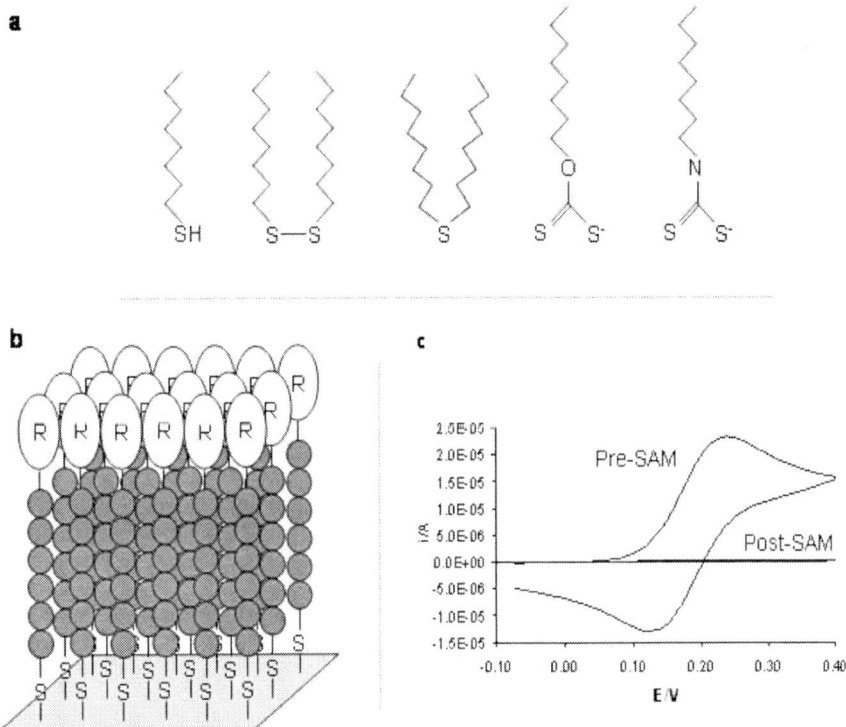

Figure 6. Surface modification using SAMs. (a) Examples of molecules that self-assemble on gold forming SAMs. From left to right: alkanethiol, dialkil sidulfide, dialkil sulfide, alkil xanthate, and dialkilthiocarbamate. (b) Illustration of the tight and uniform cover formed by SAMs over the gold surface. R indicates the free radical exhibited on surface (i.e.: amino, carboxyl, biotin, hydroxyl, maleimido, etc.). (c) Cyclic voltammogram obtained at a bare gold electrode in the presence of ferrocyanide showing the characteristic oxidation and reduction peaks. Following SAM self-assembly, the tight cover completely blocks electron transfer and ferrocyanide is not detected any more.

SAMs are often the basis for the subsequent immobilisation of biocomponents or biorecognition elements by a number of chemical conjugation or cross-binding strategies that will be described later in the text (section 3.3.4). The possibilities are endless because the functional groups provided by the SAM layer can be tailored to suit any particular requirements. Additionally, those small-size bioreceptors that exhibit as part of their structure, or can incorporate to it thiol groups, are susceptible to be directly self-assembled onto the metal surface. This option is regularly exploited for the self-assembly of aptamers and nucleic acid probes, and has been successfully applied to the immobilisation of Ab fragments. Bigger proteins can be also thiolated using the appropriate cross-linkers, followed by self-assembly on surface [17, 21, 105]. However, it will be difficult to guarantee that the construct has actually self-assembled and not just or at least partly physisorbed.

Even if a number of reports for detection of whole bacteria describe biosensors based on SAM-modification [32, 33, 106-108], we have observed unacceptable levels of bacteria non-specific adsorption onto the SAM components [15]. This is presumably related to the fact that an important number of works reporting on whole bacteria detection fail to provide the results of the appropriate negative control experiments as to demonstrate that detection is truly specific [109].

3.2.3. Silanisation of Glass and Silica Derivatives

Silanisation is a special type of self-assembly (Figure 7a). In this case, alkoxysilanes are used, which are molecules that include in their composition a silane-derived hydrolysable group (-SiX_3), typically an alkoxy group (alkyl group linked to oxygen). This group can react with the hydroxyl groups that are exposed on the surface of silica-based materials, such as glass, mica, silicon or metal oxide, forming covalent -Si-O-Si- bonds [88, 110, 111]. As silicon is one of the materials exploited in device microfabrication, silanisation is an effective and frequently used procedure for modification of chemical and physical properties of these substrates [112-114]. As happens with SAMs formed onto metal substrates, silanisation using alkoxysilanes that exhibit functional groups allows surface tailoring. The silanised surfaces can be subsequently modified by a number of well established chemical strategies in order to incorporate biocomponents (for more information, go to section

3.2.4). Interestingly, silanisation is compatible with surface micropatterning, potentially allowing modification of defined areas onto the sensor surface.

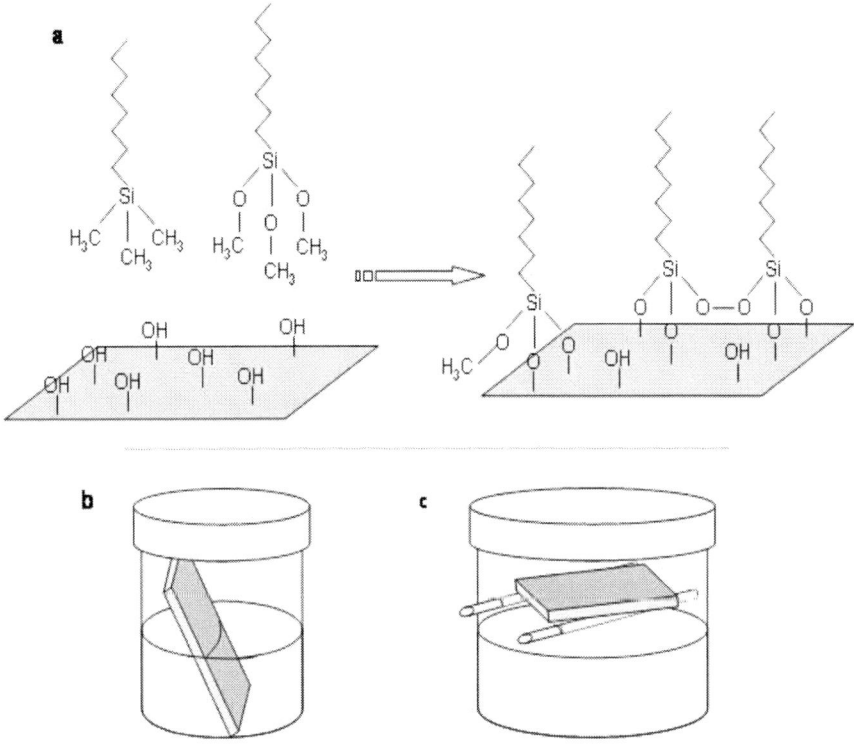

Figure 7. Surface modification by silanisation. (a) Silanisation is based on the self-assembly of silane-like molecules, mostly methylsilanes (molecule on the left) and methoxysilanes (molecule on the right) onto silicon derived materials. The -OH groups present on surface attack and displace the alkoxy groups on the silanes and form covalent -Si-O-Si- bonds. (a) Silanisation can be carried out by immersion of the physical substrate into a silane solution. Alternatively, suspension over the silane solution allows silanization by vapour deposition.

From an experimental point of view, two different deposition techniques are regularly used for silanisation. The first approach consists in surface immersion into an ethanolic solution of the alkoxysilane of choice (Figure 7b). Alternatively, the substrate can be suspended onto the alkoxysilane solution into an oxygen free sealed chamber (Figure 7c). In this case the modification takes place by vapour deposition, which prevents the formation of unnecessarily thick silane layers on the surface.

3.2.4. Chemical Conjugation

The chemical conjugation of biocomponents onto the sensor surface generates, at least in theory, stable, ordered and packed surfaces. This strategy is defended by several authors as more reproducible than random physisorption thanks to the fact that molecules are incorporated to the surface via specific reactive groups. For example, conjugation of Ab through the –COOH extreme should orient the molecule as to expose the Ab –NH_2 terminal, where target-binding takes place, and provide better orientation of the bioreceptor and more functional surfaces than random physisorption or Ab conjugation via the molecule –NH ter. In this respect, cross-binding happens between reactive groups existing on the sensor surface and *any* reactive groups exposed on the biocomponent surface, and is often coupled to SAM formation or silanisation. Hence, in fact, it is difficult to direct cross-linking towards a specific region of the biocomponent unless its structure/sequence has been well characterised (for example, reaction towards the Ab C-ter will also affect any -COOH groups existing close to the N-ter).

Chemical conjugation is generally achieved by using cross-linkers, which are chemical compounds reactive towards given groups [115]. The two most commonly used cross-linkers are glutaraldehyde and EDC (1-ethyl-3-[3-dimethylaminopropyl]carbodiimide hydrochloride). Glutaraldehyde is a highly reactive homofunctional cross-linker that provides conjugation of two amino groups (Figure 8a). "Activation" of an amine-exposing surface with glutaraldehyde makes it reactive towards amine-containing components. The reaction is extremely efficient and fast, what makes it also quite unpredictable. EDC, on the other hand, is a zero-length cross-linker that reacts with carboxyl groups forming an O-acylisourea intermediate (Figure 8b). This intermediate is reactive towards amino groups and forms very stable amide bonds. Hence EDC allows conjugation of carboxyl groups to amine groups. Because the amine reactive intermediate is unstable in aqueous solutions, the reaction is often performed in the presence of N-hydroxysuccinimide (NHS), which provides amine reactive intermediates more stable in aqueous media and enhanced reaction rates. EDC/NHS has been extensively used to "activate" SAMs and molecules exhibiting –COOH groups, so that they are reactive towards –NH_2 groups.

SMCC (Sulfosuccinimidyl 4-[N-maleimidomethyl]cyclohexane-1-carboxylate) and SATA (N-Succinimidyl-S-acetylthioacetate) are heterofunctional cross-linkers that provide conjugation between amino and thiol groups. Activation of the amino group with SMCC (or its derivatives) incorporates a highly stable maleimido group, which is reactive towards free −SH groups (Figure 9a). Thus dimmers of the activated molecule are not produced. Activation of the −NH_2 with SATA, on the other hand, generates a protected sulfhydryl group (Figure 9b). Deprotection by deacylation generates free sulfhydryl groups, which are reactive towards free −SH and S-S bonds. These components offer the additional advantage of incorporating a spacer arm between the two reactive groups. This helps to overcome steric effects which may happen when the biocomponent is immobilized too near from the surface to allow access by the receptor or appropriate folding/structure.

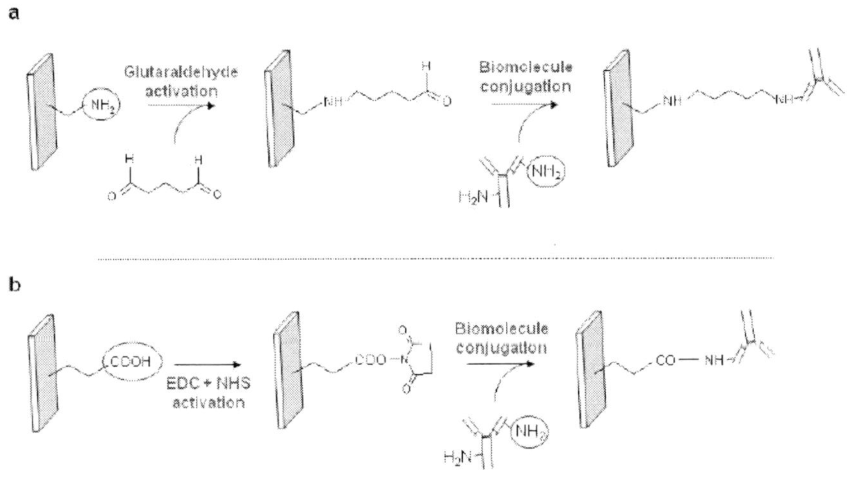

Figure 8. Surface modification by chemical modification (1). (a) Cross-linking using glutaraldehyde. The amine groups on the sensor surface are activated with glutaraldehyde. (b) Activation of the COOH groups at the sensor surface with EDC/NHS. In both cases, the surface is then reactive towards any amine groups present on the biomolecule surface.

An alternative approach for the incorporation of polysaccharide-containing components is oxidation with sodium periodate and conjugation to amine or hydrazyde groups on surface by reductive amination. This strategy has been extensively exploited for PAb modification, because conjugation through the lateral polysaccharide chains does not interfere with Ab performance. The same procedure can be used to activate surfaces exposing hydroxyl groups (such as silica or carbon), followed by conjugation of amine-containing biocomponents. Nevertheless, this protocol includes the utilization of a number of highly reactive and hazardous components, and appropriate handling and facilities are required.

The main drawback of chemical conjugation is that the highly reactive cross-linkers use to be very toxic. Their utilisation should be followed by appropriate disposal in order to avoid subsequent human or environment exposure. Chemical conjugation often involves multi-step and/or long procedures and few data are available on conjugation specificity. For example, part of the protein incubated with a chemically activated surface may just adsorb on top of it.

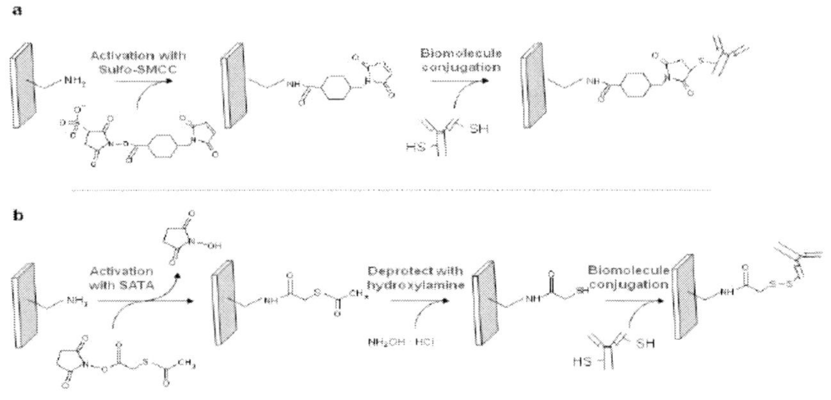

Figure 9. Surface modification by chemical modification (2). (a) Cross-linking using Sulfo-SMCC. The amine groups on the sensor surface are activated. (b) SATA incorporates a protected, thus stable, sulfhydryl group that has to be deprotected previous to conjugation. Once ready, the two surfaces are reactive towards any free sulfhydryl groups on the biomolecule surface.

3.2.5. Functionalisation by Biocomponent Entrapment

Biomolecules can be immobilised within organic or inorganic polymer matrices by entrapment during the matrix polymerisation. Entrapped molecules have been reported to survive in good shape without suffering any chemical modification that could affect their integrity. On the contrary, confinement of proteins into small inert spaces contributes to stabilise them by reducing unfolding and shifting equilibrium between different configurations [116]. For example, enzymes entrapped within the nanopores of silica beads showed improved stability for longer times than enzymes free in the bulk solution, and were applicable to biosensor development [117].

Contrary to what happens with strategies such as silanisation and SAM formation, entrapment can be carried out on a broad variety of materials. In the same manner, entrapment is possible using a variety of matrices. For example, charged molecules can be electrostatically entrapped in thermally evaporated lipid films and be later released by controlling the medium pH [118]. Cells can be entrapped into enzymatically polymerised hydrogels and yet retain growing and responding ability [119]. Enzymes entrapped within chemically polymerised hydrogels preserved their activity for storing times of up to 1 year [120]. Polyacrylamide has been used for Ab entrapment and subsequent myoglobin detection [121].

The most extensively exploited strategy for biomolecule entrapment is electrochemical entrapment from solutions containing a mixture of the bioreceptor and an electroactive monomer. When a suitable potential is applied to an electrode, the electroactive monomer polymerizes, the electropolymer grows on top of the electrode, and biomolecules are trapped within the matrix [122, 123]. Electrochemical deposition has been reported to generate reproducible and controlled film formation. On the other hand, entrapment requires high concentrations of both monomer and biomolecule, with its additional cost, and does not always generate surfaces as reproducible as desired. Additionally, entrapment can induce biomolecule loss of structure and/or function, might generate poor accessibility to certain target molecules, and not all the biomolecules are resistant to the polymerisation conditions and/or polymer components [122, 124]. As an alternative, some authors have described the polymerisation of a mixture of monomers and monomer-derivative biomolecules (i.e.: Ab, biotin), or the chemical conjugation of biomolecules to the surface of a pre-formed polymer, in an attempt to reduce the amount of biomolecule required while preserving its integrity and access to the target.

Entrapment has been mainly exploited for enzyme immobilisation followed by voltammetric or amperometric detection of its substrate in solution [125-127].

3.2.6. Bioreceptor Capture by Affinity Proteins

3.2.6.1. Protein A and Protein G

Protein A and protein G are cell surface receptors produced by several strains of *Staphylococcus aureus* and *Streptococcus sp.* respectively [128]. Both proteins have the ability to bind to Ab from several species with high efficiency (especially IgG), but in a reversible way (Table 2). For this reason, they have been extensible used in Ab isolation and purification. For example, most of the commercially available antibodies advertised by the providers as "affinity purified" have in fact been separated by using protein A/G affinity columns. Proteins A/G have also been used as capture elements in several immunochemical assay formats, including western blot, immunohistochemistry, and ELISA, as well as in immunoprecipitation assays [128].

Table 2. Comparative affinity of proteins A and G for Abs of different origin. (-) stands for no affinity and (++++) stands for high affinity

Ab from:	Ab subclass:	Binding by protein A:	Binding by protein G:
Chicken	IgY	-	-
Cow		++	++++
Dog		++	+
Goat		-	++
Hamster		+	++
Horse		++	++++
Human	IgA	variable	-
	IgD	-	-
	IgE	++	-
	IgG1	++++	++++
	IgG2	++++	++++
	IgG3	-	++++
	IgG4	++++	++++
	IgM	variable	-
Mouse	IgG1	+	++++
	IgG2a	++++	++++
	IgG2b	+++	+++
	IgG3	++	+++
	IgM1	variable	-
Pig		+++	+++
Rabbit		++++	+++
Rat	IgG1	-	+
	IgG2a	-	++++
	IgG2b	-	++
	IgG3	+	++
Sheep		+/-	++

Each molecule of protein A/G has molecular weight of approximately 40/20 KDa and can bind up to 4 and 2 molecules of IgG respectively. Protein G can also bind serum albumin; for this reason the commercial protein is usually a recombinant product from which this binding domain has been eliminated. Interestingly, proteins A/G bind the Ab through the COOH-extreme. When used on a sensor surface, this promotes highly directed capture of the Ab, which will preferentially expose the target binding site. Protein A regenerability is based in the fact that it is very resistant to extreme conditions such as heat, denaturing agents and non-ionic detergents, but its interaction with IgG is sensitive to extreme pH and salinity. Accordingly, Ab can be captured and then be efficiently eluted from protein A coated surfaces. This has lead several authors to incorporated protein A/G as a regenerable biocomponent for the development of reusable sensing surfaces [14, 129, 130]. The composition of such surfaces, however, should be carefully optimised, as any protein A/G molecule not having captured an Ab will potentially bind any Ab present in the sample under study, or any labelled Ab subsequently used for signal amplification.

3.2.6.2. The Biotin-(Strept)Avidin System

Avidin and streptavidin are two proteins that naturally exhibit high affinity for biotin (also known as vitamin H). The high specificity and low dissociation constant of the complex formed (around 10^{-15} M, one of the strongest non-covalent biological interactions known) equal those of Ab-antigen or receptor-ligand recognition. Binding, which takes place very fast, generates extremely stable complexes that remain unaffected under a wide range of pH and temperature, and in the presence of organic solvents and other denaturing agents. On top of this, biotinylation reagents for the directed modification of reactive amine, carboxylate, sulfhydryl and carbohydrate groups are commercially available and allow relatively easy biotinylation of proteins, nucleic acids and other molecules through well known chemical reactions [115, 131]. This has made (strept)avidin-biotin affinity capture one of the preferred strategies for surface functionalisation and biocomponent modification in a number of analytical and biotechnological applications [132, 133]. The main characteristics of several biotin-binding proteins have been summarized in Table 3.

Table 3. Properties and characteristics of four biotin-binding proteins. *MW*: molecular weight in molecular weight units (KDaltons); *Glycoprot*: if it is or not a glycoprotein; *Ip*: isoelectric point; *Ka*: association constant of the biotin-protein complex, in M^{-1}; *Kd*: dissociation constant of the biotin-protein complex, in mol/L; *Reg*: regenerability under mild conditions; *Binds*: number of biotin molecules bound per protein (or monomer, in brackets)

Biotin-binding protein	MW (KDa)	Glycoprot	Ip	Ka	Kd	Reg	Binds	Extracted from
Avidin	66-69	Yes	10	$\sim 10^{15}$	1.3×10^{-15}	No	4 (1)	Egg white
Neutravidin	60	No	6.3	$\sim 10^{15}$	1.3×10^{-15}	No	4 (1)	Egg white
Captavidin	--	Yes	10	$\sim 10^{9}$	--	Yes	4 (1)	Egg white
Streptavidin	53	No	5-7.5	$\sim 10^{15}$	4×10^{-14}	No	4 (1)	*Streptomyces avidinii*

Both avidin and streptavidin are tetramers formed by four identical subunits, each of them containing a binding site for biotin. Streptavidin is a 52800-Dalton biotin-binding protein produced by *Streptomyces avidinii*. Because streptavidin is a naturally non-glycosylated protein with isoelectric point, pI, around 5-6, it shows little surface charge at physiological pH and reportedly promotes low levels of biocomponent non-specific biding. Nevertheless, streptavidin dissociates 30 times faster from biotin than avidin [134]. Avidin, on the other hand, is a 66.000-Dalton glycoprotein generally obtained from egg white. In spite of its extensive utilisation in biosensor development, avidin is positively charged at neutral pH due to its high carbohydrate content (pI 10). This has been reported to promote important levels of non-specific adsorption of negatively charged components [18]. Accordingly, avidin is avoided by most researchers when whole cell bacteria or nucleic acids are targeted. This has moved providers to develop avidin derivatives of enhanced behaviour. For example, NeutrAvidinTM and ExtrAvidin®, produced by Invitrogen and Sigma respectively, are avidin-derived molecules which have been submitted to carbohydrate depletion. Hence, both components retain the high affinity characteristic of avidin but promote lower levels of non-specific adsorption than it. Neutravidin (pI 6.3) has been successfully exploited in the development of biosensors for detection of bacterial whole cells and genome [97, 135, 136].

(Strept)avidin is generally incorporated to sensing surfaces either by random physisorption, by entrapment or by chemical conjugation to preassembled SAMs or silanes. This is followed by surface blocking with, for example, an excess of unrelated protein, and by affinity capture of biotinylated bioreceptors of choice. One of the main limitations of this strategy is that it depends on the accessibility to bioreceptors which had been previously biotinylated (either commercially available or modified *in situ*), specially considering that modification of this kind usually affects negatively the receptor's performance. On the other hand, (strept)avidin affinity capture depends on multi-step modification procedures. The resulting complex surface covers often provide extremely low levels of biocomponent non-specific binding, but their thickness can also impair signal transduction efficiency of some sensing formats, particularly the electrochemical ones. These procedures are also considerably longer than other simpler alternatives, such as bioreceptor direct physisorption. Attempts to regenerate, thus reuse several times, the (strept)avidin-coated sensors by disruption of the biotin interaction generally involve harsh conditions that mostly result in (strept)avidin denaturation and permanent lost of function. A promising alternative has been described in the shape of captavidin (or its equivalent nitroavidin). Captavidin is an avidin derivative in which a tyrosine in the biotin-binding site has been nitrated [137]. As a result, captavidin binds biotin at pH 4.0-7.0 (Ka 10^9 M^{-1}), but the complex dissociates at pH 10.0. This makes captavidin a regenerable biotin-binding biocomponent, a property that has been tentatively exploited for the production of reusable sensing surfaces [138, 139]. Finally, it has to be taken into account that some proteins and tissues possess covalently bound biotin [140]. Therefore, (strept)avidin affinity capture should be avoided in certain analytical applications where the presence of endogenous biotin might generate increased levels of background signal.

3.2.7. Surface Physical Blocking and the Inclusion of Negative Controls

Independently of the strategy exploited, sensor functionalisation by incorporation of a certain bioreceptor usually generates unevenly covered surfaces. Any surface patches remaining bare will potentially promote subsequent biocomponent non-specific adsorption, contributing to increase the background signal and decrease assay specificity. This is of special importance when samples of complex matrix, as most real samples, are studied. For that reason, sensor functionalisation normally includes surface blocking steps, which consist in physical filling of any bare surface gaps with, for example, an excess on not interfering protein. By minimising non-specific adsorption of unwanted components, surface blocking contributes to guarantee that the detected targets have been specifically captured by the bioreceptors present on surface. This is crucial when reagentless sensing formats are developed, where a single capture event is directly translated into a signal and does not require subsequent steps or addition of labelled components. Typical examples include impedance and surface plasmon resonance sensors, among others (see later in the text).

The most widely used blocking agents are bovine serum albumin (BSA), casein and skimmed milk, which are commercially available at relatively low cost. Incubations of 1-2 hours at 37°C or up to 24 hours at 4°C in solutions 1-3 % (w/v) commonly generate complete surface coverage. Some non-ionic detergents, such as Tween, Triton and Nonidet P-40, have also been successfully used as blocking agents thanks to their transient effect in reducing and/or disrupting protein-protein as well as protein-surface hydrophobic interactions [141, 142]. Detergent blocking, however, has to be coupled to the use of detergent-containing solutions if all the subsequent steps, which does not always generate optimal results in the case of bacteria whole cell detection [109]. The use of polymer blocking agents, such as long-chain hydrophilic diamines, dextrans and polyethylene glycol, has been extensively reported for small target detection, specially coupled to the use of SPR sensors functionalised by self-assembly [143-145]. However, few data exist for bacteria detection and they mostly indicate that such polymers are not efficient enough on their own and/or require additional blocking with, for example, BSA [15, 30].

Finally, sensor specificity has to be assessed by studying the appropriate negative controls. Although evident, this basic premise is not always attended in the fast evolving field of biosensor development for bacteria detection. In this respect, the researcher should be able to demonstrate that any new sensor developed provides specific capture of the pathogen of choice by the bioreceptor and not just non-specific adsorption onto the physical substrate. On the other hand, the sensor should specifically capture the target pathogen but not other related microorganisms. In the first case, experiments have to be performed in the presence of the target microorganism, but using a negative control sensing surface modified with a fake non cross-reacting bioreceptor (i.e.: an Ab or a DNA probe towards a completely different target). Even if the utilisation of non-functionalised surfaces can be useful in some cases, it has to be taken into account that biomolecule incorporation modifies the surface electrochemical properties and may change its behaviour versus non-specific binding (something especially true in the case of charged molecules, such as nucleic acid probes and aptamers). In the second case, it has to be demonstrated that the sensor does not bind non-target pathogens or components potentially present in the samples under study. Also, the application of a sensor to the study of real samples will be only possible if the sample matrix does not interfere in target capture or signal transduction either.

3.3. TRANSDUCTION FORMATS USED FOR PATHOGEN DETECTION: SENSITIVITY OF THE BIOSENSOR

After binding of the analyte to the biosensor recognition element, the change in surface physical properties generated has to be converted by the transducer into a measurable signal. This can be achieved by the use of different transduction systems. According to the transduction system used, the biosensors can be divided into four main categories: electrochemical, optical, calorimetric, and piezoelectric. The selection of one or another signal transduction system depends, among others, on the expected signal type to be detected by the biosensor, the receptor and target molecule characteristics, the sample matrix we intend to study, and the assay format available [146]. The main characteristics of the different transduction systems are summarized below.

3.3.1. Electrochemical Transduction

Electrochemical biosensing is characterised by the detection of electrical changes occurring on the surface and/or the interface of mainly metal (e.g. gold, platinum) or carbon based (e.g. graphite, glassy carbon) electrodes. They are the most used so far, followed by optical biosensors. The advantages of this type of biosensors rely on the fact that they can be easily miniaturized, and low cost, easy-to-use, and portable devices can be produced, what makes possible to perform *on-site* analyses by not specially trained personnel.

Electrochemical measurements are normally carried out in an electrochemical cell (Figure 10a/b). The cell has to be big enough to fit enough solution of the appropriate conductivity as to enclose the working and reference macroelectrodes, and often also a counter electrode bigger in size than the working. This often implies an unnecessary waste of expensive reagents. Electrodes of a variety of materials, sizes, and geometries can be used (Figure 10c). The development of microfabricated microelectrodes allows the production of minute devices that readily integrate 2-3 or more electrodes (Figure 10d). Accordingly, surface functionalisation and detection are possible in extremely small volumes, down to some few microliters, decreasing the cost in reagents and volume of potentially harmful waste disposal. Microelectrodes also offer better sensitivity than their macroscopic counterparts, but their production cost is still too high for most applications and can only be produced at specialised facilities. A halfway option is the utilization of screen-printed electrodes (SPE). SPE are relatively cheap and disposable electrodes, produced by reasonably simple lithographic techniques in a variety of materials (gold, carbon, carbon nanotubes, composite materials, etc.). Multichannel formats have also been developed by integration of SPE at the bottom of the wells in microtitter plates. This allows the simultaneous monitoring of several parameters and/or samples, while taking profit of the equipment already optimised for ELISA assays [147].

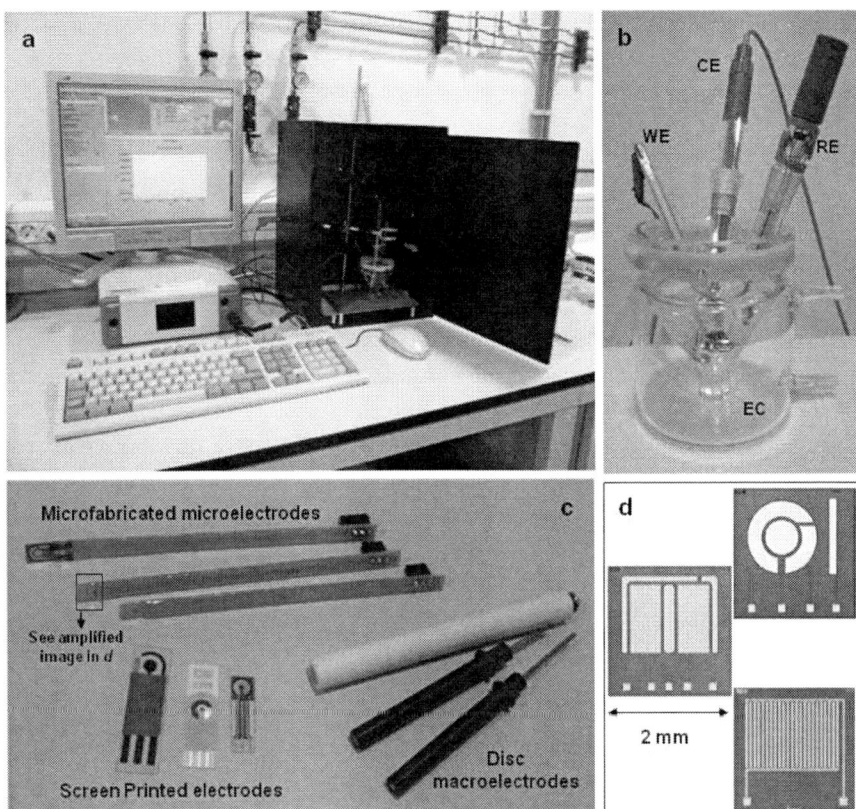

Figure 10. General setting for electrochemical measurements. (a) Example of an experimental setting showing a potentiostat connected to an electrochemical cell, were the electrochemical measure takes place, and to a computer for data acquisition and analysis. In order to prevent interferences due to environmental electrochemical noise, the reactions are often performed into a metallic Faraday's cage (the black box in the picture). (b) The electrochemical cell (EC) consists of a container filled with enough volume of conductive solution as to cover the electrodes. The CE in the picture incorporates also a flow system to facilitate temperature control over time. In most cases, three electrodes are used: working (WE), reference (RE) and counter electrode (CE). (c) A variety of working electrodes of different material, size and geometry are available. (d) shows microscope images of three examples of microfabricated microelectrodes.

The main drawback associated to electrochemical biosensing is the fact that optimal signal transduction occurs at bare electrodes. Hence, electrode functionalisation contributes to its passivation (physical blocking) and obstructs electrode transfer across it more importantly than in other

3.3.1.1. *Amperometric Biosensors*

Amperometric biosensors measure the current generated when a potential is applied between the working electrode and a reference electrode (often an Ag/AgCl reference electrode). Thus they measure movement of electrons occurring from one electrode to the other (Figure 11). Amperometric detection relies on the use of electroactive labels, which are components that can be electrochemically oxidised or reduced onto the electrode surface and generate electron transfer. In most cases, enzymes such as peroxidase, alkaline phosphatase, or glucose oxidase are used for this purpose as labels in a way very similar to ELISA. For electrochemical detection, however, an appropriate enzyme substrate has to be selected as to ensure that the enzyme will hydrolyse it into an electroactive product. For example, the enzyme β-galactosidase can hydrolyze 4-aminophenyl β-D-galactopyranoside (PAPG) into p-aminophenol (PAP), which is electroactive. When the enzymatic product is not electroactive on its own or is not efficiently detected by the system, a redox mediator has to be added to the reaction. In this case the mediator plays two roles as it guarantees both enzyme regeneration and electron shuttling between the enzyme active centre and the electrode surface. For instance, mediators such as tetramethylbenzidine (TMB) or ferrocene-derivatives, among others, are frequently added to HRP/H_2O_2 detection. Because the rate of oxidation/reduction of these redox active substrates is directly proportional to enzyme activity, they are used to monitor the electrical changes occurring after binding of the analyte to the biorecognition element. The consumption or generation of the redox substrate is easily monitored by applying to the working electrode the potential at which the redox probe is reduced or oxidized [148, 149].

Amperometric sensors are mostly based in sandwich, competition or displacement assay formats. In this context, the use of labels implies a two-binding event and improves sensor specificity. Sandwich formats, on the other hand, provide the lowest detection limits reported. Using enzyme labels also offers the possibility to generate signal amplification by extending the enzymatic reaction over time. On the contrary, the dependence on the use of labels implies the performance on long and multi-step assays, with numerous washings in-between them, making difficult the integration in lab-on-chip devices.

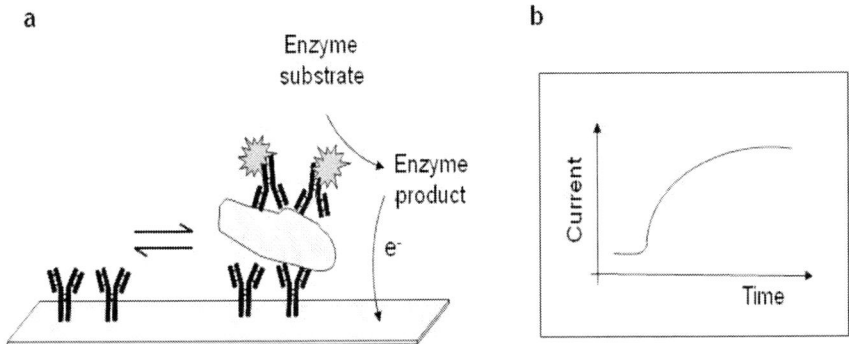

Figure 11. Amperometric biosensors. Amperometric biosensors measure the current generated over time due to accumulation of an electroactive label. They are mostly exploited in sandwich assay formats, where a first biocomponent is immobilised onto the sensor surface and behaves as a capture element of the target, while a second enzyme-labelled biocomponent works as the reporter element. Enzyme activity is proportional to the concentration of target captured.

3.3.1.2. Impedimetric Biosensors

Impedimetric biosensors are based on the electrochemical impedance spectroscopy technique (EIS), which is a powerful method capable of detecting small changes occurring at either the bulk solution or onto the electrode surface (Figure 12). The system parameters most commonly measured are the changes in capacitance and electron transfer resistance occurring at the electrode-solution interface, and the variation in medium conductivity. This combination of the analysis of both resistive and capacitive properties allows the reagentless and real-time study of the biocomponents deposited on the microelectrode surface, and therefore is highly useful for the detection of binding events [150].

Briefly, IES is based on the perturbation of a system at equilibrium with a small-amplitude sinusoidal current at different frequencies. Given the complexity of real systems, impedance data are frequently analysed according to simplified models. For example, the obtained impedance spectra can be modelled and fitted by using a system's equivalent circuit, in which each element is expressed in terms of electrical components such as resistances and capacitances, for the analysis of the electrical properties of the microelectrode interface [5]. Such data analysis is normally carried out using a software program that performs fitting analysis of the different parameters according to the pre-defined equivalent circuit. However, it is also possible to simplify the measurement and the analysis by reducing the number of the frequencies assayed to a few or even a single one for biosensor applications [82, 150].

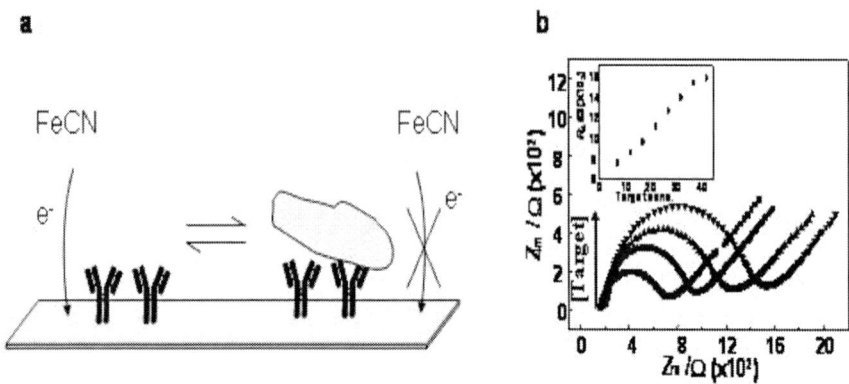

Figure 12. Impedimetric biosensors. Impedance is a reagentless transducing format. Thus target binding is directly detected due to the changes induced in surface electrochemical behaviour. (a) Capture of a target molecule/microorganism on surface induces in most cases an increase in electrode transfer resistance from the solution to the electrode surface. Some exceptions are the study of bacterial metabolism over time and bacteria capture using lytic phages, which correlate with the release of high concentrations of electroactive intracellular components. (b) Results can be analysed and represented in a number of ways. For example, Nyquist plots show a semicircle section, which diameter gives information about electron transfer resistance, and a lineal segment, indicative of the level of system dependence on mass transfer.

Impedance is a label free technique, what makes it very attractive for biosensor applications. The equipment required is compact, with no mobile parts, and easy to miniaturize. Besides, EIS provides very low detection limits, depending on the format, with some works reporting on the detection of as low as 10^3 CFU/mL [151]. One of its main drawbacks, on the other hand, is the complexity of data analysis, especially in the study of real samples where adsorption of matrix components onto the sensor surface can generate false positive results and/or interfere with signal transduction.

3.3.1.3. Conductimetric Biosensors

Conductimetric biosensors detect conductance changes in the medium between two electrodes due to changes in the concentration of charged particles or components. For example, as a result of the bacteria metabolism some uncharged substrates present in the medium used for growing the bacteria, like carbohydrates, are transformed into different charged intermediates, like lactic acid, and can then be detected with the use of a potentiostat [152]. Such changes in medium conductance can be correlated with bacteria growth. This approach, however, does not provide target specificity beyond the specificity granted by the culture media.

3.3.1.4. Potentiometric Biosensors

Potentiometric biosensors measure the difference in potential (voltage) occurring between the working electrode and a reference electrode when no current is being flowed between them. The changes in potential detected are expected to be due to the change in the concentration of certain ions on the surface of the electrode. As to ensure so, the working electrode incorporates a semi-permeable membrane that allows certain ions to reach the electrode selectively [2].

In the context of bacteria detection, potentiometric biosensors can target an ion product of the cellular metabolism [153]. In most of the cases, however, they detect ions produced by an enzymatic label. In this approach, sandwich assay formats have been mainly reported [154, 155].

3.3.2. Optical Transduction

The optical transducers measure changes in the light properties (e.g. luminescence, fluorescence emission, refractive index) after reaching the sensor surface, where the analyte-biorecognition element interaction takes place. Together with the electrochemical sensors, optical transducers have been the most frequently used transducers for the detection of microorganisms. Different approaches have been successfully developed for the detection of the different properties of the light, including fibre-optic sensors, microarrays, surface plasmon resonance (SPR), and evanescence wave sensors. While fibre-optic and microarray biosensors require labelling with a luminescent or fluorescent marker, SPR and some evanescence wave guide biosensors offer the advantage of being a label free detection method. The main characteristics of each method are discussed below.

3.3.2.1. Fiber-Optic Biosensors

Biosensors based on this technique require immobilisation of the biorecognition element onto the core of an optical fibre, coupled to the use of fluorescent labels. Label binding produces fluorescent emission, which is directly correlated with the amount of target captured by the receptor on surface. The change in signal is thus proportional to the analyte concentration present in the sample [156]. The incident and returned light can be monitored in real-time with a specific photo-detector.

In most cases, detection at fiber-optic biosensors is based in sandwich assay formats, where a fluorescently-labelled detector Ab is added, but competitive and displacement approaches can also be optimised [157, 158]. Alternatively, the fluorescent label can be directly incorporated to the biorecognition element. In this specific setting, target capture interferes with light emission by impairing steric hindrance. One of the main advantages of fiber-optic sensors is that small low-cost laser diodes are available as excitation sources. In addition, a high number of fluorophores (and fluorophore labelled biocomponents), characterised by different excitation and emission wavelengths, are commercially available. This makes possible the optimisation of multiplex sensing formats by using in parallel biocomponents against different targets labelled with different fluorophores. On the opposite, a wide variety of components induce fluorophore quenching and their presence in samples can generate false negative results.

3.3.2.2. Microarray Biosensors

Microarrays are in most cases fluorescent biosensors in which the number of targets examined at a time is enormously increased. The way to do so is to use substrates (usually specially treated glass slides) where up to thousands of capture biocomponents are printed or spotted in parallel. Detection takes advantage of the high number of fluorophores available, as well as the physical separation between the different markers on surface. In this way, a single incubation with the sample under study allows to analyse the presence/absence of a number of targets (Figure 13).

Microarrays have been broadly used for nucleic acid hybridisation, often coupled to the study of PCR products, and in some cases allowing detection of single mismatch pairings and point mutations. Examples of protein arrays have also been reported and are being rapidly incorporated to proteomics. To the best of our knowledge, no reports exist for the detection of whole-cell microorganisms. The requirement for the use of fluorescent labels conditions the assay formats that can be exploited, which are mostly sandwich and competition approaches. Furthermore, a number of commercial array spotters, array analysers, and ready to use array kits can be purchased.

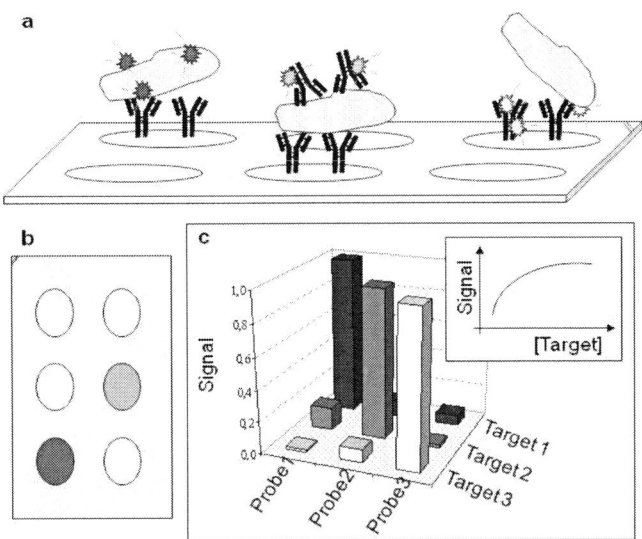

Figure 13. Microarray biosensors. (a & b) Microarrays consist of glass slides where a number of biorecognition elements have been separately spotted. Coupled to fluorescent detection in a variety of assay formats, microarrays allow easy and fast multiplexed detection of up to hundreds target ligands in a sample. (a, left) competition assay; (a, middle) sandwich assay; (a, right) direct labelling of the bioreceptor immobilised on surface. (c) Signal analysis can be performed in real time and simultaneously for all the spots.

3.3.2.3. Surface Plasmon Resonance (SPR)

SPR is a physical phenomenon whereby visible light can be coupled into the electron field (plasmon wave) in a thin metal film (usually gold or silver) to create an energy wave that is guided by the metal–dielectric interface [159]. This implies that part of the light is somehow converted into vibration of the metal molecules. The energy wave generated decays exponentially in its way into the dielectric and is affected by any changes occurring in close proximity of the metal surface.

SPR chips consist of a slide made of optical glass or quartz, covered with a thin layer of gold (usually 50 nm). A transitional layer of chromium is generally added to promote gold adhesion. In most commercial SPR instruments, a mobile glass prism is used to couple laser light into the metal film surface at different incidence angles (Figure 14a/b) [160]. The equipment measures in each case the light reflected. At a certain angle of incidence of the incident light with the metal-prism interface, maximum coupling of the light with the surface plasmon occurs (Figure 14c/d). This is the angle of resonance and correlates with a drop in the amount of light reflected. The angle of resonance depends on the dielectric constant of the dielectric on the prism. Hence, binding of an analyte onto the metal surface causes a change in its dielectric constant, which changes the angle of resonance (Figure 14d). This change in resonance angle can be followed in real time, thus providing kinetic information on film formation and/or target capture [161]. However, it has to be remembered that signal transduction will be also affected by any external parameters able to influence the solution optical properties, such as temperature. For that reason, the experimental conditions have to be accurately controlled in order to provide reproducible SPR results.

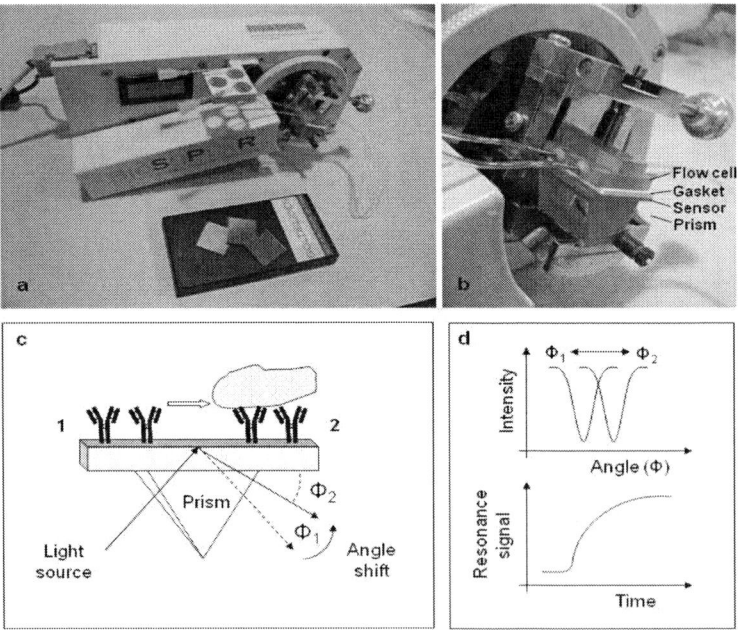

Figure 14. SPR biosensors. SPR is one of the most sensitive reagentless transducing formats. Its application to bacteria detection, however, is limited by their big size, which places most of the cell too far away from the sensing surface. (a) Example of a portable SPR equipment with some SPR chips in front of it. (b) Detail of the mobile prism, with the assembly of an SPR sensor, silicon gasket and flow cell on top. (c) Illustration of SPR performance. The angle at which the incident light beam is reflected suffers a change when the sensor surface is modified (for example, by capture of target molecules/microorganisms by the bioreceptor). (d) Signal monitoring can be performed in real time either by studying the angle of resonance (upper graph) or the resonance signal (lower graph).

SPR has generated limited results for whole bacteria detection, when the signals are usually significantly lower than expected compared to the signals measured for smaller targets. For example, most SPR sensors reported for bacteria detection show LODs above 10^5 CFU/mL. This is due to the fact that the evanescent wave formed under SPR conditions decays as it penetrates the solution and usually shows a maximal penetration of about 300 nm. Therefore, only the refractive index changes occurring closer from the sensor surface than 300 nm will cause a signal change. If we consider the average size of, for example, an *E. coli* cell (2–6 x 0.7–1.5 µm), we will easily understand that only a small portion of the captured bacteria is close enough to the sensor surface as to contribute to producing a response signal [162]. As it will be exposed later in the text, both the study of cell lysates and the development of modified measurement equipment have been explored in order to overcome this obstacle.

3.3.2.4. Evanescence Waveguide Biosensors

Waveguide biosensing is based in the physical property by which a light beam hitting the interface between two transparent media (e.g. the surface of a transparent sensor and a sample solution) at an angle of incidence greater than the *critical angle* suffers total internal reflection. The evanescent wave generated penetrates the reached material, i.e. the surface of the sensor, an order of a wavelength beyond the reflecting surface. However, the propagation of the optical wave through the sensor will be affected by any changes occurring at the sensor-media interface. For example, optical attenuation can be caused by target binding onto the sensor surface if it induces a variation in either the refractive index or the absorption coefficient. Most evanescent waveguide sensors monitor the changes occurred in the coupling angle of a laser beam into a planar waveguide, in which an optical grating has been pierced, caused by the changes in the refractive index generated upon binding of a biomolecule to the surface [163].

3.3.3. Calorimetric Transduction

Calorimetric biosensors exploit the inherent characteristics of biological reactions, which result in absorption or release of heat in the surrounding system [164]. These changes can be monitored by the use of thermometer-like sensors. Calorimetric biosensors have been extensively used for the detection of catalytic reactions carried out by enzymes immobilized on a substrate (see

[165] for review). This transduction system has not been exploited for the detection of bacteria because of its limitations of heat loss by convection, conduction or irradiation.

3.3.4. Mass-Sensitive and Resonant Sensors

Resonance sensing is based in the propagation of a mechanical oscillation (an acoustic wave) across or along the sensor/resonator surface [166]. The analyzer registers in this sensing format the mechanical resonant frequency over time and converts it into electrical oscillation frequency for its subsequent analysis. The resonant frequency depends on the velocity of the travelling wave and on the characteristics of the sensor (geometric dimensions, material properties, surface composition). Increase in the sensor mass load, for example following target capture by the ligand immobilised on it, directly translates into a change in the resonance frequency measured, which will be proportional to the mass of the molecular species bound. Thus resonators are reagent-less or label-less sensors. Accordingly, their specificity absolutely depends on the absence of cross-sensitivity and/or non-specific binding of unwanted components. Luckily, this transduction format is often compatible with the incorporation of relatively thick layers onto the sensor surface. This advantage has been exploited for the optimization of functionalisation approaches by incorporating surface blocking strategies and/or components, which have provided extraordinary low levels of non-specific adsorption but would impair signal transduction in other sensing formats. On the other hand, the sensitivity of a resonant sensor will depend on the ratio of mass change to the overall vibrating mass. For this reason, lighter resonators will generate larger relative mass increases induced by binding of a certain amount of target. The field has taken advantage of the recent advances in microfabrication technologies that have lead towards the fabrication of smaller/lighter devices and even sensor arrays, whilst making possible the integration of electronic circuitry and mass production at low cost. Apart of providing an extremely fast detection strategy, resonant sensing is well-suited for the study of optically opaque or viscous samples. Its main drawback is the limited sensitivity, often well below other sensors such as the optical (especially SPR) and electrochemical ones.

3.3.4.1. Quartz Crystal Resonators and Quartz Crystal Microbalance

The starting point of resonance sensing can be tracked back to 1959, when Sauerbrey published the essential relationship between the change of the resonant frequency of a quartz crystal and the mass added to its surface [167]. The sensor that originated from this work was called quartz crystal microbalance (QCM) and was used as an analytical tool for the first time in 1964 [168]. Since then, QCM has been extensively used to monitor film thickness (growth) in vacuum-deposition units, as well as for sensing purposes [169].

A QCM sensor consists of a quartz crystal with a gold electrode deposited on top of its surface. The application of an alternating voltage of the right frequency to the electrode generates a mechanical oscillation in the quartz crystal. The resonant frequency of this wave depends on the sensor mass loading and will suffer a shift following target capture (Figure 15). On top of this, the gold electrode provides a surface onto which the biorecognition element can be immobilised because numerous biofunctionalisation protocols have been successfully optimised for this material. Nevertheless, it has to be beared in mind that, similarly to other biosensors, the QCM frequency response is influenced, not only by mass charge, but also by a number of factors including effective viscosity, conductivity, dielectric constant, electrode morphology, density, and temperature of the solution. Hence, experimental conditions have to be accurately optimised and controlled in order to generate reproducible results and prevent undesired interferences.

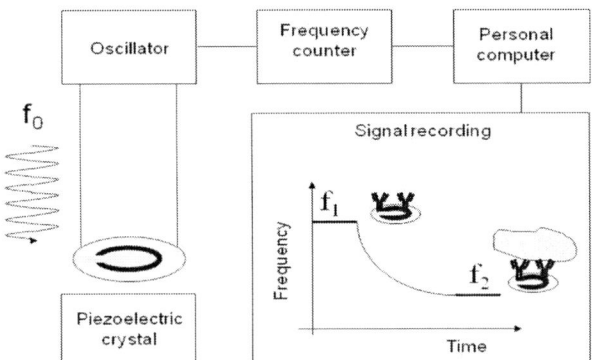

Figure 15. QCM biosensors. Piezoelectric transduction works by inducing sensor physical resonation and monitoring its frequency over time. Target binding induces changes on the sensor load, which translate into changes in the resonation frequency.

Bacteria detection by QCM is rapid and does not require target labelling or addition of labelled reagents. The procedure would optimally require only sample addition and incubation. However, QCM has seldom showed sensitivity below 10^4 CFU/mL, with most works reporting LODs between 10^5 and 10^7 CFU/mL. This is apparently due to he fact that most of the reported QCM immunosensors solely measure changes in resonant frequency, which are usually explained by the so-called Sauerbrey equation. However, several authors have exposed that the Sauerbrey equation is applicable only to thin (approximately 1μm) and elastic films coupled to the crystal surface, where the mass loading correspond to up to 0.05% of the crystal mass [170]. Accordingly, the Sauerbrey equation would not apply to the case of cells attached to the QCM surface, mainly due to their softness and relatively large size. In this respect, Su & Li proposed that the simultaneous study of resonant frequency and the viscoelastic properties of the bound surface mass (by monitoring motional resistance) would provide more accurate results [171]. Nevertheless, their results do not seem in accordance with this proposal and they report in their work LODs in the range 10^5-10^6 CFU/mL for *Salmonella typhimurium*. In a different way, Shen et al, proposed that the low levels of detectability reported for big targets on QCM sensors were mainly caused by capture via flexible interactions [172]. In order to demonstrate this, they carried out bacterial entrapment using simultaneously a carbohydrate on surface and a lectin in solution as the recognition biocomponents. The sensors generated stronger adhesion of *E. coli* cells to the surface, increased contact area between them, and improved detectability, with an assay linear range between 7.5×10^2 and 7.5×10^7 cells/mL, compared to LODs above 10^7 CFU/mL for bacteria capture via just carbohydrate on surface.

The main drawback of QCM application to bacteria detection is, as happens with other mechanical sensors, that QCM optimal performance is obtained when measuring in air instead of by immersion. This implies that the measurement can not be performed on-line and that part of the cell's mass is lost due to dehydration. Nevertheless, some recent examples have demonstrated that the measures can also be performed in aqueous solutions using microfluidic channels [173].

3.3.4.2. Surface Acoustic Wave (SAW) Sensors

In SAW sensing, an acoustic wave is generated by a piezoelectric material, for example quartz, which converts an oscillating electric field into a mechanical wave. The acoustic wave is then propagated along or across the surface of the sensor, usually an interdigitated microstructure deposited on top of the piezoelectric. The output signal is converted again by the transducer into an electrical signal for its analysis. Subsequent changes in the sensor surface loading translate into decrease of the velocity of propagation of the wave. This can result in reduction of the resonance frequency of the resonator or in modification of the phase shift between the output and input signals [166].

The design of SAW sensors is compatible with sensor miniaturization using photolithography microfabrication. This allows mass and low-cost production, as well as the production of wire-less platforms of extremely low power consumption. As happens with QCM, most of the published examples are based on measurement in non-liquid environments but an increasing number of reports describe detection modes by sample immersion.

3.3.4.3. Cantilevers

Cantilevers are composed by a beam or bar, suspended in the air by one of its ends, and are extensively used in Atomic Force Microscopy [47, 174-178]. Two main types of cantilevers are used with biosensing purposes [179]. Resonant cantilevers work very similarly to QCM: target capture induces increase in mass load onto the sensor and drop of its resonant frequency. This is the mode most frequently applied to the study of liquid samples. On the other hand, non-resonant cantilevers are mostly based in detection of the stress bending generated by the ligand-target capture event. For example, target capture at the tip of the cantilever results in physical/chemical interaction with the components of this part of the surface cover. This induces surface stress and translates into differential physical bending of the sensor (Figure 16).

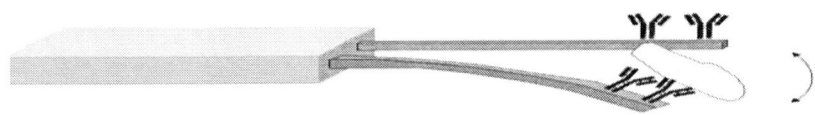

Figure 16. Cantilever biosensors. Cantilevers consist of suspended beams of minute size. When used in non-resonant mode, target capture induces surface stress which can translate into physical bending.

As already exposed, the detection limits of the resonators can be considerably improved by decreasing their size. Cantilevers have proven to be particularly effective in this regard, and microfabricated cantilevers of a great variety of dimensions and shapes, usually made of silicon nitride, covered or not with a metal layer, have been reported.

REFERENCES

[1] J. Homola (2008). Surface plasmon resonance sensors for detection of chemical and biological species. *Chemical Reviews*, *108*, 462-493.

[2] D. Ivnitski, I. Abdel-Hamid, P. Atanasov and E. Wilkins (1999). Biosensors for detection of pathogenic bacteria. *Biosensors and Bioelectronics*, *14* 599-624.

[3] O. Lazcka, F. J. Del Campo and F. X. Munoz (2007). Pathogen detection: A perspective of traditional methods and biosensors. *Biosensors and Bioelectronics*, *22*, 1205-1217.

[4] B. Pejcic, R. De Marco and G. Parkinson (2006). The role of biosensors in the detection of emerging infectious diseases. *Analyst*, *131*, 1079-1090.

[5] L. Yang and R. Bashir (2008). Electrical/electrochemical impedance for rapid detection of foodborne pathogenic bacteria. *Biotechnology Advances*, *26*, 135-150.

[6] C. García-Aljaro and Baldrich (2009). Recent advances in pathogen detection methods. In A. P. V. E. Lutsenko, *Water Microbiology: Types, Analyses and Disease-Causing Microorganisms.*, Nova Science Publishers, Inc.

[7] M. Alvarez-Icaza and U. Bilitewski (1993). Mass production of biosensors. *Analytical Chemistry*, *65*, 525A-533A.

[8] R. D. Schmid and F. E. Scheller Biosensors: application in medicine, environmental protection, and process control. New York, USA: Weinheim; 1989.

[9] A. P. F. Turner, I. Karube and G. S. Wilson *Biosensors. Fundamentals and applications*. Oxford: Oxford University Press; 1987.

[10] J. Wang (2006). Electrochemical biosensors: Towards point-of-care cancer diagnostics. *Biosensors and Bioelectronics*, *21*, 1887-1892.

[11] A. M. Azevedo, D. M. Prazeres, J. M. Cabral and L. P. Fonseca (2005). Ethanol biosensors based on alcohol oxidase. *Biosensors and Bioelectronics*, *21*, 235-247.

[12] J. D. Newman and S. J. Setford (2006). Enzymatic biosensors. *Molecular Biotechnology*, *32*, 249-268.

[13] J. Wang (2008). Electrochemical glucose biosensors. *Chemical Reviews*, *108*, 814-825.

[14] G. P. Anderson, M. A. Jacoby, F. S. Ligler and K. D. King (1997). Effectiveness of protein A for antibody immobilization for a fiber optic biosensor. *Biosensors and Bioelectronics 12*, 329-336

[15] E. Baldrich, O. Laczka, F. J. del Campo and F. X. Munoz (2008). Self-assembled monolayers as a base for immunofunctionalisation: unequal performance for protein and bacteria detection. *Analytical and Bioanalytical Chemistry*, *390*, 1557-1562.

[16] N. K. Chaki and K. Vijayamohanan (2002). Self-assembled monolayers as a tunable platform for biosensor applications *Biosensors and Bioelectronics*, *17*, 1-12.

[17] J. M. Fowler, M. C. Stuart and D. K. Y. Wong (2007). Self-assembled layer of thiolated protein G as an immunosensor scaffold. *Analytical Chemistry*, *79*, 350-354.

[18] N. M. Green (1990). Avidin and streptavidin. *Methods Enzymology*, *184*, 51-67.

[19] A. Karyakin, G. Presnova, M. Rubtsova and A. Egorov (2000). Oriented immobilization of antibodies onto the gold surfaces via their native thiol groups. *Analytical Chemistry*, *72*, 3805-3811.

[20] O. Laczka, E. Baldrich, F. J. del Campo and F. X. Munoz (2008). Immunofunctionalisation of gold transducers for bacterial detection by physisorption. *Analytical and Bioanalytical Chemistry*, *391*, 2825-2835.

[21] I.-S. Park and N. Kim (1998). Thiolated *Salmonella* antibody immobilization onto the gold surface of piezoelectric quartz crystal. *Biosensors and Bioelectronics 13*, 1091-1087.

[22] I. Vikholm (2005). Self-assembly of antibody fragments and polymers onto gold for immunosensing. *Sensors and Actuators, B: Chemical*, *106*, 311-316.

[23] A. F. Collings and F. Caruso (1997). Biosensors: recent advances. *Reports on Progress in Physics*, *60*, 1397-1445.

[24] R. S. Yalow and S. A. Berson (1959). Assay of plasma insulin in human subjects by immunological methods. *Nature*, *184*, 1648-1649.

[25] G. Kohler and C. Milstein (1975). Continuous cultures of fussed cells secreting antibody of predefined specificity. *Nature 256*, 495-497.

[26] P. Holliger and P. J. Hudson (2005). Engineered antibody fragments and the rise of single domains. *Nature Biotechnology*, *23*, 1126-1136.

[27] H. R. Hoogenboom (2005). Selecting and screening recombinant antibody libraries. *Nature Biotechnology*, *23*, 1105-1116.

[28] D. Saerens, L. Huang, K. Bonroy and S. Muyldermans (2008). Antibody fragments as probe in biosensor development. *Sensors*, *8*, 4669-4686.

[29] J. C. O'Brien, V. W. Jones, M. D. Porter, C. L. Mosher and E. Henderson (2000). Immunosensing platforms using spontaneously

adsorbed antibody fragments on gold. *Analytical Chemistry*, 72, 703-710.

[30] E. Baldrich, O. Laczka, F. J. Del Campo and F. X. Munoz (2008). Gold immuno-functionalisation via self-assembled monolayers: Study of critical parameters and comparative performance for protein and bacteria detection. *Journal of Immunological Methods*, 336, 203-212.

[31] T. T. Huang, J. Sturgis, R. Gomez, T. Geng, R. Bashir, A. K. Bhunia, J. P. Robinson and M. R. Ladisch (2003). Composite surface for blocking bacterial adsorption on protein biochips. *Biotechnology and Bioengineering*, 81, 618-624.

[32] B. K. Oh, Y. K. Kim, K. W. Park, W. H. Lee and J. W. Choi (2004). Surface plasmon resonance immunosensor for the detection of *Salmonella typhimurium*. *Biosensors and Bioelectronics*, 19, 1497-1504.

[33] A. D. Taylor, J. Ladd, Q. M. Yu, S. F. Chen, J. Homola and S. Y. Jiang (2006). Quantitative and simultaneous detection of four foodborne bacterial pathogens with a multi-channel SPR sensor. *Biosensors and Bioelectronics*, 22, 752-758.

[34] J. W. Waswa, C. Debroy and J. Irundayaraj (2006). Rapid detection of *Salmonella enteritidis* and *Escherichia coli* using Surface Plasmon Resonance Biosensor. *Journal of Food Process Engineering*, 29 373-385.

[35] S. Balasubramanian, I. B. Sorokulova, V. J. Vodyanoy and A. L. Simonian (2007). Lytic phage as a specific and selective probe for detection of *Staphylococcus aureus* - A surface plasmon resonance spectroscopic study. *Biosensors and Bioelectronics*, 22, 948-955.

[36] R. S. Lakshmanan, R. Guntupalli, J. Hu, D.-J. Kim, V. A. Petrenko, J. M. Barbaree and B. A. Chin (2007). Phage immobilized magnetoelastic sensor for the detection of *Salmonella typhimurium*. *Journal of Microbiological Methods*, 71, 55-60.

[37] L. Gervals, M. Gel, B. Allain, M. Tolba, L. Brovko, M. Zourob, R. Mandeville, M. Griffiths and S. Evoy (2007). Immobilization of biotinylated bacteriophages on biosensor surfaces. *Sensors and Actuators, B: Chemical*, 125, 615-621.

[38] R. Edgar, M. McKinstry, J. Hwang, A. B. Oppenheim, R. A. Fekete, G. Giulian, C. Merril, K. Nagashima and S. Adhya (2006). High-sensitivity bacterial detection using biotin-tagged phage and quantum-dot nanocomplexes. *Proceedings of the National Academy of Sciences of the United States of America*, 103, 4841-4845.

[39] G. P. Smith (1985). Filamentous fusion phage - novel expression vectors that display cloned antigens on the virion surface. *Science*, *228*, 1315-1317.

[40] A. Rasooly and K. E. Herold (2006). Biosensors for the analysis of food- and waterborne pathogens and their toxins. *Journal of Aoac International*, *89*, 873-883.

[41] E. R. Goldman, M. P. Pazirandeh, J. M. Mauro, K. D. King, J. C. Frey and G. P. Anderson (2000). Phage-displayed peptides as biosensor reagents. *Journal of Molecular Recognition*, *13*, 382-387.

[42] R. Jelinek and S. Kolusheva (2004). Carbohydrate biosensors. *Chemical Reviews*, *104*, 5987-6015.

[43] F. Kauffmann (1947). The serology of the coli group. *Journal of Immunology*, *57*, 71-100.

[44] K. L. Hsu, K. T. Pilobello and L. K. Mahal (2006). Analyzing the dynamic bacterial glycome with a lectin microarray approach. *Nature Chemical Biology*, *2*, 153-157.

[45] B. Serra, M. Gamella, A. J. Reviejo and J. M. Pingarrón (2008). Lectin-modified piezoelectric biosensors for bacteria recognition and quantification. *Analytical and Bioanalytical Chemistry*, *391*, 1853-1860.

[46] O. A. Sadik and F. Yan (2007). Electrochemical biosensors for monitoring the recognition of glycoprotein-lectin interactions. *Analytical Chimica Acta*, *588*, 292-296.

[47] Z. Shen, M. Huang, C. Xiao, Y. Zhang, X. Zeng and P. G. Wang (2007). Nonlabeled quartz crystal microbalance biosensor for bacterial detection using carbohydrate and lectin recognitions. *Analytical Chemistry*, *79*, 2312-2319.

[48] Q. Lu, H. Lin, S. Ge, S. Luo, Q. Cai and C. A. Grimes (2009). Wireless, remote-query, and high sensitivity *Escherichia coli* O157:H7 biosensor based on the recognition action of concanavalin A. *Analytical Chemistry*, *81*, 5846-5850.

[49] S. Arcidiacono, P. Pivarnik, C. M. Mello and A. Senecal (2008). Cy5 labeled antimicrobial peptides for enhanced detection of *Escherichia coli* O157 : H7. *Biosensors & Bioelectronics*, *23*, 1721-1727.

[50] N. V. Kulagina, G. P. Anderson, F. S. Ligler, K. M. Shaffer and C. R. Taitt (2007). Antimicrobial peptides: New recognition molecules for detecting botulinum toxins. *Sensors*, *7*, 2808-2824.

[51] J. E. Dover, G. M. Hwang, E. H. Mullen, B. C. Prorok and S.-J. Suh (2009). Recent advances in peptide probe-based biosensors for detection of infectious agents. *Journal of Microbiological Methods*, *78*, 10-19.

[52] S. R. Mikkelsen (1996). Electrochemical biosensors for DNA sequence detection. *Electroanalysis*, *8*, 15-19.
[53] F. R. R. Teles and L. R. Fonseca (2008). Trends in DNA biosensors. *Talanta*, *77*, 606-623.
[54] A. Sassolas, B. D. Leca-Bouvier and L. J. Blum (2008). DNA biosensors and microarrays. *Chemical Reviews*, *108*, 109-139.
[55] S. Cosnier (2005). Affinity biosensors based on electropolymerized films. *Electroanalysis*, *17*, 1701-1715.
[56] H. Peng, C. Soeller, M. B. Cannell, G. A. Bowmaker, R. P. Cooney and J. Travas-Sejdic (2006). Electrochemical detection of DNA hybridization amplified by nanoparticles. *Biosensors and Bioelectronics*, *21*, 1727-1736.
[57] R. H. Liu, J. N. Yang, R. Lenigk, J. Bonanno and P. Grodzinski (2004). Self-contained, fully integrated biochip for sample preparation, polymerase chain reaction amplification, and DNA microarray detection. *Analytical Chemistry*, *76*, 1824-1831.
[58] X. L. Mao, L. J. Yang, X. L. Su and Y. B. Li (2006). A nanoparticle amplification based quartz crystal microbalance DNA sensor for detection of *Escherichia coli* O157 : H7. *Biosensors and Bioelectronics*, *21*, 1178-1185.
[59] B. Elsholz, R. Worl, L. Blohm, J. Albers, H. Feucht, T. Grunwald, B. Jurgen, T. Schweder and R. Hintsche (2006). Automated detection and quantitation of bacterial RNA by using electrical microarrays. *Analytical Chemistry*, *78*, 4794-4802.
[60] F. Farabullini, F. Lucarelli, I. Palchetti, G. Marrazza and M. Mascini (2006). Disposable electrochemical genosensor for the simultaneous analysis of different bacterial food contaminants. *Biosensors and Bioelectronics 22*, 1544-1549
[61] E. Katz and I. Willner (2003). Probing biomolecular interactions at conductive and semiconductive surfaces by impedance spectroscopy: Routes to impedimetric immunosensors, DNA-Sensors, and enzyme biosensors. *Electroanalysis*, *15*, 913-947.
[62] M. Egholm, O. Buchardt, L. Christensen, C. Behrens, S. M. Freier, D. A. Driver, R. H. Berg, S. K. Kim, B. Norden and P. E. Nielsen (1993). PNA hybridizes to complementary oligonucleotides obeying the Watson-Crick hydrogen-bonding rules. *Nature*, *365*, 566-568.
[63] P. E. Nielsen, M. Egholm, R. H. Berg and O. Buchardt (1991). Sequence-selective recognition of DNA by strand displacement with a thymine-substituted polyamide. *Science*, *254*, 1497-1500.

[64] K. K. Jensen, H. Orum, P. E. Nielsen and B. Norden (1997). Kinetics for hybridization of peptide nucleic acids (PNA) with DNA and RNA studied with the BIAcore technique. *Biochemistry*, 36, 5072-5077.

[65] J. Wang (1998). DNA biosensors based on peptide nucleic acid (PNA) recognition layers. A review. *Biosensors and Bioelectronics*, 13, 757-762.

[66] J. Wang, P. E. Nielsen, M. Jiang, X. Cai, J. R. Fernandes, D. H. Grant, M. Ozsoz, A. Beglieter and M. Mowat (1997). Mismatch-sensitive hybridization detection by peptide nucleic acids immobilized on a quartz crystal microbalance. *Analytical Chemistry*, 69, 5200-5202.

[67] O. Brandt and J. D. Hoheisel (2004). Peptide nucleic acids on microarrays and other biosensors. *Trends in Biotechnology*, 22, 617-622.

[68] H. A. Joung, N. R. Lee, S. K. Lee, J. Ahn, Y. B. Shin, H. S. Choi, C. S. Lee, S. Kim and M. G. Kim (2008). High sensitivity detection of 16s rRNA using peptide nucleic acid probes and a surface plasmon resonance biosensor. *Analytica Chimica Acta*, 630, 168-173.

[69] S. Reisberg, L. A. Dang, Q. A. Nguyen, B. Piro, V. Noel, P. E. Nielsen, L. A. Le and M. C. Pham (2008). Label-free DNA electrochemical sensor based on a PNA-functionalized conductive polymer. *Talanta*, 76, 206-210.

[70] A. Cattani-Scholz, D. Pedone, F. Blobner, G. Abstreiter, J. Schwartz, M. Tornow and L. Andruzzi (2009). PNA-PEG Modified Silicon Platforms as Functional Bio-Interfaces for Applications in DNA Microarrays and Biosensors. *Biomacromolecules*, 10, 489-496.

[71] A. D. Ellington and J. W. Szostak (1990). In vitro selection of RNA molecules that bind specific ligands. *Nature*, 346, 818-822.

[72] D. L. Robertson and G. F. Joyce (1990). Selection in vitro of an RNA enzyme that specifically cleaves single-stranded DNA. *Nature*, 344, 467-468.

[73] C. Tuerk and L. Gold (1990). Systematic evolution of ligands by exponential enrichment: RNA ligands to bacteriophage T4 DNA polymerase. *Science*, 249, 505-510.

[74] E. Torres-Chavolla and E. C. Alocilja (2009). Aptasensors for detection of microbial and viral pathogens. *Biosensors and Bioelectronics*, 24, 3175-3182.

[75] S. Klussmann, A. Nolte, R. Bald, V. A. Erdmann and J. P. Fürste (1996). Mirror-image RNA that binds D-adenosine. *Nature Biotecnology*, 14, 1112-1116.

[76] A. Nolte, S. Klussmann, R. Bald, V. A. Erdmann and J. P. Fürste (1996). Mirror-design of L-oligonucleotide ligands binding to L-arginine. *Nature Biotechnology*, *14*, 1116-1119.

[77] W. Kusser (2000). Chemically modified nucleic acid aptamers for in vitro selections: evolving evolution. *Journal of Biotechnology*, *74*, 27-38.

[78] F. J. Del Campo, O. Ordeig, N. Vigues, N. Godino, J. Mas and F. X. Munoz (2007). Continuous measurement of acute toxicity in water using a solid state microrespirometer. *Sensors and Actuators, B: Chemical*, *126*, 515-521.

[79] P. Melidis, E. Vaiopoulou and A. Aivasidis (2008). Development and implementation of microbial sensors for efficient process control in wastewater treatment plants. *Bioprocess and Biosystems Engineering*, *31*, 277-282.

[80] E. Z. Ron (2007). Biosensing environmental pollution. *Current Opinion in Biotechnology*, *18*, 252-256.

[81] C. Garcia-Aljaro, X. Munoz-Berbel, A. T. Jenkins, A. R. Blanch and F. X. Munoz (2008). Surface plasmon resonance assay for real-time monitoring of somatic coliphages in wastewaters. *Applied and Environmental Microbiology*, *74*, 4054-4058.

[82] X. Muñoz-Berbel, C. García-Aljaro and F. J. Muñoz (2008). Impedimetric approach for monitoring the formation of biofilms on metallic surfaces and the subsequent application to the detection of bacteriophages. *Electrochimica Acta*, *53*, 5739-5744.

[83] C. Garcia-Aljaro, X. Munoz-Berbel and F. J. Munoz (2009). On-chip impedimetric detection of bacteriophages in dairy samples. *Biosensors and Bioelectronics*, *24*, 1712-1716.

[84] R. J. Ansell (2001). MIP-ligand binding assays (pseudo-immunoassays). *Bioseparation*, *10*, 365-377.

[85] O. Hayden, P. A. Lieberzeit, D. Blaas and F. L. Dickert (2006). Artificial antibodies for bioanalyte detection-sensing viruses and proteins. *Advanced Functional Materials*, *16*, 1269-1278.

[86] S. D. Harvey, G. M. Mong, R. M. Ozanich, J. S. McLean, S. M. Goodwin, N. B. Valentine and J. K. Fredrickson (2006). Preparation and evaluation of spore-specific affinity-augmented bio-imprinted beads. *Analytical and Bioanalytical Chemistry*, *386*, 211-219.

[87] A. Namvar and K. Warriner (2007). Microbial imprinted polypyrrole/poly(3-methylthiophene) composite films for the detection of *Bacillus* endospores. *Biosensors and Bioelectronics*, *22*, 2018-2024.

[88] S. K. Parida, S. Dash, S. Patel and B. K. Mishra (2006). Adsorption of organic molecules on silica surface. *Advances in Colloid and Interface Science*, *121*, 77-110.

[89] F. Heitz and N. Van Mau (2002). Protein structural changes induced by their uptake at interfaces *Biochimica et Biophysica Acta 1597*, 1-11.

[90] J. E. Butler (2000). Solid supports in enzyme-linked immunosorbent assay and other solid-phase immunoassays. *Methods*, *22*, 4-23.

[91] Z. P. Chen, J. H. Jiang, G. L. Shen and R. Q. Yu (2005). Impedance immunosensor based on receptor protein adsorbed directly on porous gold film. *Analytica Chimica Acta*, *553*, 190-195.

[92] Z.-P. Chen, Z.-F. Peng, P. Zhang, X.-F. Jin, J.-H. Jiang, X.-B. Zhang, G.-L. Shen and R.-Q. Yu (2007). A sensitive immunosensor using colloidal gold as electrochemical label. *Talanta 72*, 1800-1804.

[93] M. Kobayashi, M. Sato, Y. Li, N. Soh, K. Nakano, K. Toko, N. Miura, K. Matsumoto, A. Hemmi, Y. Asano and T. Imato (2005). Flow immunoassay of trinitrophenol based on a surface plasmon resonance sensor using a one-pot immunoreaction with a high molecular weight conjugate. *Talanta*, *68*, 198-206.

[94] R. M. Pemberton, T. T. Mottram and J. P. Hart (2005). Development of a screen-printed carbon electrochemical immunosensor for picomolar concentrations of estradiol in human serum extracts. *Journal of Biochemical and Biophysical Methods*, *63*, 201-212.

[95] M. K. Sharma, A. K. Goel, L. Singh and V. K. Rao (2006). Immunological biosensor for detection of *Vibrio cholerae* O1 in environmental water samples. *World Journal of Microbiology and Biotechnology*, *22*, 1155-1159.

[96] G. Volpe, G. Fares, F. d. Quadri, R. Draisci, G. Ferretti, C. Marchiafava, D. Moscone and G. Palleschi (2006). A disposable immunosensor for detection of 17β-estradiol in non-extracted bovine serum. *Analytica Chimica Acta*, *572*, 11-16.

[97] O. Laczka, E. Baldrich, F. X. Munoz and F. J. del Campo (2008). Detection of *Escherichia coli* and *Salmonella typhimurium* using interdigitated microelectrode capacitive immunosensors: The importance of transducer geometry. *Analytical Chemistry*, *80*, 7239-7247.

[98] J. C. Love, L. A. Estroff, Kriebel, J.K., R. G. Nuzzo and G. M. Whitesides (2005). Self-assembled monolayers of thiolates on metals as a form of nanotechnology. *Chemical Reviews*, *105*, 1103-1169.

[99] C. D. Bain and G. M. Whitesides (1988). Formation of two-component surfaces by the spontaneous assembly of monolayers on gold from

solutions containing mixtures of organic thiols. *Journal of the American Chemical Society*, *110*, 6560-6561.

[100] R. G. Nuzzo and D. L. Allara (1983). Adsorption of bifunctional organic disulfides on gold surfaces. *Journal of the American Chemical Society*, *105*, 4481-4483.

[101] E. B. Troughton, C. D. Bain, G. M. Whitesides, R. G. Nuzzo, D. L. Allara and M. D. Porter (1988). Monolayer films prepared by the spontaneous self-assembly of symmetrical and unsymmetrical dialkyl sulfides from solution onto gold substrates: Structure, properties, and reactivity of constituent functional groups. *Langmuir*, *4*, 365-385.

[102] V. M. Mirsky (2002). New electroanalytical applications of self-assembled monolayers. *Trends in Analytical Chemistry*, *21*, 439-450.

[103] T. Wink, S. J. van Zuilen, A. Bult and W. P. van Bennkom (1997). Self-assembled monolayers for biosensor. *Analyst*, *122*, 43R-50R.

[104] J. Hautman and M. L. Klein (1990). Molecular dynamics simulation of the effects of temperature on a dense monolayer of long-chain molecules. *The Journal of Chemical Physics*, *93*, 7483-7492.

[105] J. C. Pyun, S. D. Kim and J. W. Chung (2005). New immobilization method for immunoaffinity biosensors by using thiolated proteins. *Analytical Biochemistry*, *347*, 227-233.

[106] J. Y. Jyoung, S. H. Hong, W. Lee and J. W. Choi (2006). Immunosensor for the detection of *Vibrio cholerae* O1 using surface plasmon resonance. *Biosensors and Bioelectronics*, *21*, 2315-2319.

[107] X. L. Su and Y. B. Li (2004). A self-assembled monolayer-based piezoelectric immunosensor for rapid detection of *Escherichia coli* O157 : H7. *Biosensors and Bioelectronics*, *19*, 563-574.

[108] A. Subramanian, J. Irudayaraj and T. Ryan (2006). A mixed self-assembled monolayer-based surface plasmon immunosensor for detection of *E-coli* O157 : H7. *Biosensors and Bioelectronics*, *21*, 998-1006.

[109] E. Baldrich, N. Vigues, J. Mas and F. X. Munoz (2008). Sensing bacteria but treating them well: Determination of optimal incubation and storage conditions. *Analytical Biochemistry*, *383*, 68-75.

[110] Y. Duvault, A. Gagnaire, F. Gardies, N. Jaffrezic-Renault, C. Martelet, D. Morel, J. Serpinet and J.-L. Duvault (1990). Physicochemical characterization of covalently bonded alkyl monolayers on silica surfaces. *Thin Solid Films*, *185*, 169-179.

[111] E. Ruckenstein and Z. F. Li (2005). Surface modification and functionalization through the self-assembled monolayer and graft polymerization. *Advances in Colloid and Interface Science*, *113*, 43-63.

[112] S. K. Bhatia, L. C. Shriver-Lake, K. J. Prior, J. H. Georger, J. M. Calvert, R. Bredehorst and F. S. Ligler (1989). Use of thiol-terminal silanes and heterobifunctional crosslinkers for immobilization of antibodies on silica surfaces. *Analytical Biochemistry*, *178*, 408-413.

[113] K. Saal, T. Tätte, I. Tulp, I. Kink, A. Kurg, U. Mäeorg, A. Rinken and A. Lõhmus (2006). Sol-gel films for DNA microarray applications. *Materials Letters*, *60*, 1833-1838.

[114] N. M. Pope, D. L. Kulcinski, A. Hardwick and Y.-A. Chang (1993). New application of silane coupling agents for covalently binding antibodies to glass and cellulose solid supports. *Bioconjugate Chemistry*, *4*, 166-171.

[115] G. T. Hermanson *Bioconjugate techniques*: Academic press, London; 1996.

[116] H.-X. Zhou and K. A. Dill (2001). Stabilization of Proteins in Confined Spaces. *Biochemistry*, *40*, 11289-11293.

[117] S. Sotiropoulou, V. Vamvakaki and N. A. Chaniotakis (2005). Stabilization of enzymes in nanoporous materials for biosensor applications. *Biosensors and Bioelectronics*, *20*, 1674-1679.

[118] M. Sastry (2002). Entrapment of proteins and DNA in thermally evaporated lipid films. *Trends in Biotechnology 20*, 185-188.

[119] A. D. Taylor, J. Ladd, Q. M. Yu, S. F. Chen, J. Homola and S. Y. Jiang (2006). Quantitative and simultaneous detection of four foodborne bacterial pathogens with a multi-channel SPR sensor. *Biosensors and Bioelectronics*, *22*, 752-758.

[120] A. Guiseppi-Elie, N. F. Sheppard Jr., S. Brahim and D. Narinesingh (2001). Enzyme microgels in packed-bed bioreactors with downstream amperometric detection using microfabricated interdigitated microsensor electrode arrays. *Biotechnology and Bioengineering*, *75*, 475-484.

[121] C. M. Hanbury, W. G. Miller and R. B. Harris (1997). Enzyme microgels in packed-bed bioreactors with downstream amperometric detection using microfabricated interdigitated microsensor electrode arrays. *Clinical Chemistry*, *43*, 2128-2136.

[122] S. Cosnier (2003). Biosensors based on electropolymerized films: new trends. *Analytical and Bioanalytical Chemistry*, *377*, 507-520.

[123] S. Cosnier (1999). Biomolecule immobilisation on electrode surfaces by entrapment or attachment to electrochemically polymerized films. *Biosensors and Bioelectronics, 14*, 443-456

[124] A. I. Minett, J. N. Barisci and G. G. Wallace (2002). Coupling conducting polymers and mediated electrochemical responses for the detection of *Listeria*. *Analytica Chimica Acta, 475*, 37-45.

[125] P. N. Bartlett, P. R. Birkin, J. H. Wang, F. Palmisano and G. De Benedetto (1998). An Enzyme Switch Employing Direct Electrochemical Communication between Horseradish Peroxidase and a Poly(aniline) Film. *Analytical Chemistry 70*, 3685-3694

[126] D. Shan, Y. Y. He, S. X. Wang, H. G. Xue and H. Zheng (2006). A porous poly(acrylonitrile-co-acrylic acid) film-based glucose biosensor constructed by electrochemical entrapment. *Analytical Biochemistry, 356*, 215-221.

[127] H. B. Yidiz and L. Toppare (2006). Biosensing approach for alcohol determination using immobilized alcohol oxidase. *Biosensors and Bioelectronics, 21*, 2306-2310.

[128] J. Goding (1978). Use of staphylococcal protein A as an immunological reagent. *Journal of Immunological Methods, 20*, 241-253.

[129] S. Babacan, P. Pivarnik, S. Letcher and A. Rand (2000). Evaluation of antibody immobilization methods for piezoelectric biosensor application. *Biosensors and Bioelectronics, 15*, 615-621.

[130] T. O'Brien, L. H. Johnson, J. L. Aldrich, S. G. Allen, L. T. Liang, A. L. Plummer, S. J. Krak and A. A. Boiarski (2000). The development of immunoassays to four biological threat agents in a bidiffractive grating biosensor. *Biosensors and Bioelectronics, 14*, 815-828.

[131] M. D. Savage, G. Mattson, S. Desai, G. W. Nielander, S. Morgensen and E. J. Conklin *Avidin-Biotin Chemistry: A handbook*. Rockford, Illinois: Pierce Chemical Cmpany.; 1992.

[132] M. Wilchek and E. A. Bayer (1988). The avidin-biotin complex in bioanalytical applications. *Analytical Biochemistry, 171*, 1-32.

[133] O. H. Laitinen, H. R. Nordlund, V. P. Hytonen and M. S. Kulomaa (2007). Brave new (strept)avidins in biotechnology. *Trends in Biotechnology, 25*, 269-277.

[134] U. Piran and W. J. Riordan (1990). Dissociation rate constant of the biotin-streptavidin complex. *Journal of Immunological Methods, 133*, 141-143.

[135] A. K. Deisingh and M. Thompson (2001). Sequences of *E. coli* O157:H7 detected by a PCR-acoustic wave sensor combination. *Analyst, 126*, 2153-2158.

[136] H. Sun, Y. Zhang and Y. Fung (2006). Flow analysis coupled with PQC/DNA biosensor for assay of *E. coli* based on detecting DNA products from PCR amplification. *Biosensors and Bioelectronics, 22*, 506-512.

[137] E. Morag, E. A. Bayer and M. Wilchek (1996). Reversibility of biotin-binding by selective modification of tyrosine in avidin. *Biochemical Journal, 316*, 193-199.

[138] J. G. Bolivar, S. A. Soper and R. L. McCarley (2008). Nitroavidin as a Ligand for the Surface Capture and Release of Biotinylated Proteins. *Analytical Chemistry, 80*, 9336-9342.

[139] C. García-Aljaro, F. X. Muñoz and E. Baldrich (2009). Captavidin: a new regenerable biocomponent for biosensing? *Analyst, 134*, 2338–2343.

[140] G. S. Wood and R. Warnke (1981). Suppression of endogenous avidin-binding activity in tissues and its relevance to biotin-avidin detection systems. *Journal of Histochemistry and Cytochemistry, 29*, 1196-1204.

[141] G. E. Kenny and C. L. Dunsmoor (1987). Effectiveness of detergents in blocking nonspecific binding of IgG in the enzyme-linked immunosorbent assay (ELISA) depends upon the type of polystyrene used. *Israel Journal of Medical Sciences, 23*, 732-734.

[142] M. Steinitz (2000). Quantitation of the blocking effect of tween 20 and bovine serum albumin in ELISA microwells. *Analytical Biochemistry 282*, 232-238.

[143] W. Knoll, M. Zizlsperger, T. Liebermann, S. Arnold, A. Badia, M. Liley, D. Piscevic, F.-J. Schmitt and J. Spinke (2000). Streptavidin arrays as supramolecular architectures in surface-plasmon optical sensor formats. *Colloids and Surfaces A: Physicochemical and Engineering Aspects, 161*, 115-137.

[144] B. Johnsson, S. Lofas and G. Lindquist (1991). Immobilization of Proteins to a Carboxymethyldextran-Modified Gold Surface for Biospecific Interaction Analysis in Surface Plasmon Resonance Sensors. *Analytical Biochemistry, 198*, 268-277.

[145] D. J. O'Shannessy, M. Brigham-Burke and K. Peck (1992). Immobilization chemistries suitable for use in the BIAcore surface plasmon resonance detector. *Analytical Biochemistry, 205*, 132-136.

[146] M. Gerard, A. Chaubey and B. D. Malhotra (2002). Application of conducting polymers to biosensors. *Biosensors and Bioelectronics*, *17*, 345-359.

[147] O. Bagel, C. Degrand, B. Limoges, M. Joannes, F. Azek and P. Brossier (2000). Enzyme affinity assays involving a single-use electrochemical sensor. Applications to the enzyme immunoassay of human chorionic gonadotropin hormone and nucleic acid hybridization of human cytomegalovirus DNA. *Electroanalysis*, *12*, 1447-1452.

[148] P. Leonard, S. Hearty, J. Brennan, L. Dunne, J. Quinn, T. Chakraborty and R. O'Kennedy (2003). Advances in biosensors for detection of pathogens in food and water. *Enzyme and Microbial Technology*, *32*, 3-13.

[149] P. D. Patel (2002). (Bio)sensors for measurement of analytes implicated in food safety: a review. *Trac-Trends in Analytical Chemistry*, *21*, 96-115.

[150] O. Panke, T. Balkenhohl, J. Kafka, D. Schafer and F. Lisdat (2008). Impedance spectroscopy and biosensing. *Advances in Biochemical Engineering/Biotechnology*, *109*, 195-237.

[151] C. Ruan, L. Yang and Y. Li (2002). Immunobiosensor chips for detection of *Escherichia coli* O157:H7 using electrochemical impedance spectroscopy. *Analytical Chemistry 74* 4814-4820.

[152] E. R. Richter (1993). Biosensors: applications for dairy food industry. *Journal of Dairy Science*, *76*, 3114-3117.

[153] D. C. Cowell, A. A. Dowman, R. J. Lewis, R. Pirzad and S. D. Watkins (1994). The rapid potentiometric detection of catalase positive microorganisms. *Biosensors and Bioelectronics*, *9*, 131-138.

[154] [154]   C. Ercole, M. del Gallo, M. Pantalone, S. Santucci, L. Mosiello, C. Laconi and A. Lepidi (2002). A biosensor for *Escherichia coli* based on a potentiometric alternating biosensing (PAB) transducer. *Sensors and Actuators, B: Chemical*, *83*, 48-52.

[155] K. A. Uithoven, J. C. Schmidt and M. E. Ballman (2000). Rapid identification of biological warfare agents using an instrument employing a light addressable potentiometric sensor and a flow-through immunofiltration-enzyme assay system. *Biosensors and Bioelectronics*, *14*, 761-770.

[156] A. Leung, P. M. Shankar and R. Mutharasan (2007). A review of fiber-optic biosensors. *Sensors and Actuators, B: Chemical*, *125*, 688-703.

[157] T. Geng, J. Uknalis, S.-I. Tu and A. K. Bhunia (2006). Fiber-optic biosensor employing alexa-fluor conjugated antibody for detection of

Escherichia coli O157:H7 from ground beef in four hours. *Sensors*, *6*, 796-807.

[158] L. C. Shriver-Lake, J. P. Golden, G. Patonay, N. Narayanan and F. S. Ligler (1995). Use of three longer-wavelength fluorophores with the fiber-optic biosensor. *Sensors and Actuators, B: Chemical*, *29*, 25-30.

[159] A. J. C. Tubb, F. P. Payne, R. B. Millington and C. R. Lowe (1997). Single-mode optical fibre surface plasma wave chemical sensor. *Sensors and Actuators, B: Chemical*, *41*, 71-79.

[160] E. Kretschmann (1971). Determination of optical constants of metals by excitation of surface plasmons. *Zeitschrift für Physik*, *241*, 313.

[161] B. Liedberg, C. Nylander and I. Lundstrom (1995). Biosensing with surface plasmon resonance--how it all started. *Biosensors and Bioelectronics*, *10*, i-ix.

[162] P. M. Fratamico, T. P. Strobaugh, M. B. Medina and A. G. Gehring (1998). Detection of *Escherichia coli* O157 : H7 using a surface plasmon resonance biosensor. *Biotechnology Techniques*, *12*, 571-576.

[163] K. Tiefenthaler (1993). Grating couplers as label-free biochemical waveguide sensors. *Biosensors and Bioelectronics*, *8*, R35-R37.

[164] C. Spink and I. Wadso (1976). Calorimetry as an analytical tool in biochemistry and biology. *Methods of Biochemical Analysis*, *23*, 1-159.

[165] K. Ramanathan and B. Danielsson (2001). Principles and applications of thermal biosensors. *Biosensors and Bioelectronics*, *16*, 417-423.

[166] R. Lucklum and P. Hauptmann (2006). Acoustic microsensors-the challenge behind microgravimetry. *Analytical and Bioanalytical Chemistry*, *384*, 667-682.

[167] G. Sauerbrey (1959). Verwendung von Schwingquarzen zur Wägung dünner Schichten und zur Mikrowägung. *Zeitschrift Physik*, *155*, 206-212.

[168] W. H. King Jr (1964). Piezoelectric sorption detector. *Analytical Chemistry*, *36*, 1735-1739.

[169] M. A. Cooper and V. T. Singleton (2007). A survey of the 2001 to 2005 quartz crystal microbalance biosensor literature: Applications of acoustic physics to the analysis of biomolecular interactions. *Journal of Molecular Recognition*, *20*, 154-184.

[170] C. Lu and A. W. Czanderna *Applications of Piezoelectric Quartz Crystal Microbalances*. Amsterdam, New York: Elsevier; 1984.

[171] X. L. Su and Y. B. Li (2005). A QCM immunosensor for *Salmonella* detection with simultaneous measurements of resonant frequency and motional resistance. *Biosensors and Bioelectronics*, *21*, 840-848.

[172] Z. H. Shen, M. C. Huang, C. D. Xiao, Y. Zhang, X. Q. Zeng and P. G. Wang (2007). Nonlabeled quartz crystal microbalance biosensor for bacterial detection using carbohydrate and lectin recognitions. *Analytical Chemistry*, *79*, 2312-2319.

[173] B. Godber, K. S. J. Thompson, M. Rehak, Y. Uludag, S. Kelling, A. Sleptsov, M. Frogley and M. A. Cooper (2005). Direct quantification of analyte concentration by resonant acoustic profiling. *Clinical Chemistry 51*, 1962-1972.

[174] K. M. Goeders, J. S. Colton and L. A. Bottomley (2008). Microcantilevers: Sensing chemical interactions via mechanical motion. *Chemical Reviews*, *108*, 522-542.

[175] K. M. Hansen and T. Thundat (2005). Microcantilever biosensors. *Methods*, *37* 57-64.

[176] R. Raiteri, M. Grattarola, H. J. Butt and P. Skládal (2001). Micromechanical cantilever-based biosensors. *Sensors and Actuators, B: Chemical*, *79*, 115-126.

[177] A. Raman, J. Melcher and R. Tung (2008). Cantilever dynamics in atomic force microscopy. *Nano Today*, *3*, 20-27.

[178] C. Ziegler (2004). Cantilever-based biosensors. *Analytical and Bioanalytical Chemistry*, *379*, 946-959.

[179] R. Mutharasan (2008). Cantilever Sensors for Pathogen Detection. Principles of Bacterial Detection: Biosensors, Recognition Receptors and Microsystems, 459-480, In: Zourub et al. (Eds). *Principles of Bacterial Detection: Biosensors, Recognition Receptors and Microsystems*, Springer, N.Y., 1012 pp.

Chapter 4

BIOSENSORS FOR PATHOGEN DETECTION

The biosensors for pathogen detection developed to date have mainly focused on the detection of the most common waterborne and foodborne pathogens: *E. coli* (especially the serotype O157:H7), *Campylobacter*, *Cryptosporidium*, *Legionella*, *Listeria*, and *Salmonella*. Tables 4, 5 and 6 summarize the different biosensor formats reported and their application to the detection of different pathogens. It is worth stressing that, although the number of studies is noteworthy, the majority of the studies published describe laboratory tests performed in PBS or similar solutions rather than in real samples. In addition, a significant number of publications in the field do not present the appropriate negative controls to demonstrate that the binding event detected is truly specific and not caused by, for example, non-specific adsorption of the sample components onto the sensor surface [1]. Even if apparently trivial, this kind of considerations illustrate why the number of publications appeared does not necessarily correlate with development of commercial devices.

Electrochemical biosensing has shown very promising in the field of bacteria detection, as it provides assay format versatility, exceptionally low detection limits and the possibility to develop truly hand held-miniaturized biosensors. Accordingly, an important proportion of the works published report on the electrochemical detection of either bacteria whole cells or bacterial cell components. An interesting approximation entails the detection of the activity of enzymes characteristic of the bacterial metabolism, which takes advantage of the fact that most of them involve in fact electrochemical reduction/oxidation reactions. The exploitation of other common transducing formats, such as SPR or FET, has been limited for whole cell bacteria detection. This is mainly due to the target big size, which places most of it outside the effective sensing distance for this kind of sensors. Some few works report on bacteria successful detection using alternative transducing devices, such as cantilevers or QCM, or taking advantage of novel materials including carbon nanotubes and nanowires.

Some examples of biosensors reported for the detection of different microorganism types will be described along the diverse sections of this chapter. We have selected those reporting either the most original approaches, or the best limits of detection. In addition, we have priorized those works showing appropriate data validation. For example, the majority of the works selected include negative controls to confirm assay specificity and/or have been performed in real sample matrices.

4.1. DETECTION OF BACTERIAL METABOLISM AND/OR BACTERIAL ENZYMES

The *ATP bioluminescence assay* was first applied to measure microorganisms in food back in the 1970s [2]. The assay was based on the fact that ATP is a major biological energy source which is present in most bacteria at quite a constant concentration (approximately 0.47 fg per cell) and has been formatted over the last decades into a number of commercially available assay kits of extremely simple handling and high efficiency [3]. The procedure typically includes hydrolysis of any ATP free in solution, release of bacterial ATP, and reaction with the firefly enzyme *luciferin* to produce bioluminescence. The intensity of the bioluminescence emitted is then directly related to the amount of living microorganisms potentially present. Luo and co-workers recently reported the formatting of the classical ATP bioluminescence assay into a sensing format [4]. They coupled a chemical method for ATP extraction from the bacterial cells to fluorescence detection using a homemade sensor and luminometer. The system was able to detect *E. coli* and *S. aureus* at a concentration range from 10^3 to 10^8 CFU/mL in food samples in just 5 minutes.

The fact that a number of biological reactions are in fact oxidations or reductions, and/or produce electroactive products, has been exploited for the optimisation of a number of unique sensing approaches based on the amperometric monitoring of bacterial metabolism. Detection of a metabolic product provides, among others, assay low-cost and simplicity, the possibility to target viable cells, and a direct detection approach that does not require subsequent addition of labels or labelled components and is often compatible with signal amplification strategies. Nevertheless, and although extremely sensitive, this type of biosensor suffers of limited specificity that is often counterbalanced by sample pre-enrichment in selective culture media. This is the basis of *Impedance Microbiology*, the simplest electrochemical sensing approach for bacteria detection, and the only one that has been commercialized to date. Impedance microbiology consists in the monitoring over time of the impedance registered at a bare electrode, which has been immersed in a sample potentially containing bacteria. Impedance is an indicator of electron transfer efficiency across the solution-electrode interface and is affected by both modification of the electrode surface and changes in the solution conductivity. The method is based on the fact that bacterial metabolism results in the conversion of complex molecules present in the sample or culture medium, such as proteins and polysaccharides, into ionic metabolites, such as aminoacids and organic acids [5-7]. The appearance of such components contributes to modify the medium conductivity, and finally affects the impedance component of the electrode. When combined with traditional microbiology (i.e.: sample pre-enrichment for 24-48-hours in selective media) impedance microbiology may provide identification of different microorganism groups (ranging from total aerobic plate count to coliforms, yeast, mold, or lactic acid bacteria) to even specific pathogens (such as *E. coli*, *Salmonella* or *Pseudomonas*). This is the base of a number of existing commercial impedimetric systems, including the Bactometer® from Biomerieux; BacTrac® from Sy-Lab, Malthus® System from Malthus Instruments, and RABIT® from Don Whitley Scientific. Alternative attempts to provide impedance microbiology with specificity have also been reported. For example, Yang and Li incorporated a bacteria capture and pre-concentration step using immuno-magnetic particles previous to impedance monitoring [8]. This allowed them to detect *Salmonella* in a concentration range from 10^1 to 10^6 CFU/mL, following resuspension in Brain Heart Infusion Medium and incubation for 1.5 to 8 hours, with no interference from other microbial species such as *E. coli*.

A related approximation is the biosensor described by Morales and co-workers for the detection of *E. coli*, which is based on the amperometric detection of glucose consumption during bacterial growth [9]. This biosensor showed a limit of detection of 6.5 x 10^2 and 6.5 CFU/mL after 3 and 7 h of incubation, respectively, for *E. coli*. This method, in combination with the addition of antibiotics to the growth medium, allowed the specific identification of a number of different microorganisms, including *E. coli*, *Staphylococcus aureus* and *Salmonella choleraesuis*, on the basis of their different susceptibility to the antibiotics tested.

The detection of β-galactosidase as a bacterial faecal indicator to monitor faecal pollution has been extensively exploited as an alternative to direct pathogen detection. β-galactosidase is an enzyme which is expressed by all the bacteria in the coliform group and thus leads to detection of total colifoms. The approach addressed by Serra and co-workers [10] consisted in a biosensor for the indirect detection of β-galactosidase activity (Figure 17). The procedure included 5 hours of sample enrichment at 37°C, induction of the enzyme with IPTG, and permeabilisation of the cells in order to release all the molecules of β-galactosidase produced. This was followed by addition of a suitable enzyme substrate, phenyl β-D-galactopyranoside, which is hydrolyzed by β-galactosidase into amino-phenol. Because amino-phenol is an electroactive molecule, it could be then amperometrically detected at the surface of a novel graphite-teflon-tyrosinase composite electrode. The sensor allowed detection down to 10 CFU/mL *E. coli*. The limit of detection could be further decreased to 1 CFU/100mL, which is the law requirement for coliform bacteria in drinking water, after pre-concentration of one litter of sample and less than 6.5 hours of enrichment. Cheng and co-workers coupled this detection strategy to the utilisation of a tyrosinase / Fe_3O_4 magnetic nanoparticle / carbon nanotube nanocomposite biosensor and detection by flow injection assay (FIA) [11]. They succeeded to detect *E. coli*, spiked in water and then filtered and resuspended in LB media supplemented with IPTG, in a concentration range between 20 and 10^5 CFU/mL. The limit of detection was of 10 CFU/mL in an assay of about 4 hours. However, the procedure included a number of sample filtration steps that hardened manipulation. A certain level of handling simplification was reported in the work by Boyaci, who coupled amperometric detection of β-galactosidase to bacteria pre-concentration using antibody-functionalised magnetic particles, followed by incubation at 37°C in culture media supplemented with enzyme inducer and enzyme substrate [12]. The authors reported detection of 2×10^6 CFU/mL of *E. coli* after 30 minutes of incubation and total analysis time of less than 1 h. Detectability could be additionally improved down to 20 CFU/mL *E. coli* if sample enrichment was extended to 6–7 hours.

Attempts to provide specificity to β-galactosidase sensing have been reported by Rishpon's team [13]. In this work, the authors performed sample pre-enrichment at 37°C in culture media supplemented with IPTG and a lytic phage (λ). This was followed by filtration in order to eliminate undisrupted whole cells and cell debris. The bacteria-depleted filtrate was then supplemented with enzyme substrate and generation of amino-phenol was amperometrically monitored over time using disposable carbon screen printed electrodes. The authors claimed detection of as low as 1 *E. coli* CFU in a 100 mL sample within 6-8 hours of incubation. Significantly lower signals were obtained both in *E. coli* incubated in the absence of phages and in the presence of phage and *Klebsiella pneumoniae*, another β-galactosidase producing bacteria that is not infected by λ. This format was later formatted for detection of alpha-glucosidase and beta-glucosidase, which are enzymes constitutively expressed by *Bacillus cereus* and *Mycobacterium smegmatis* respectively [14].

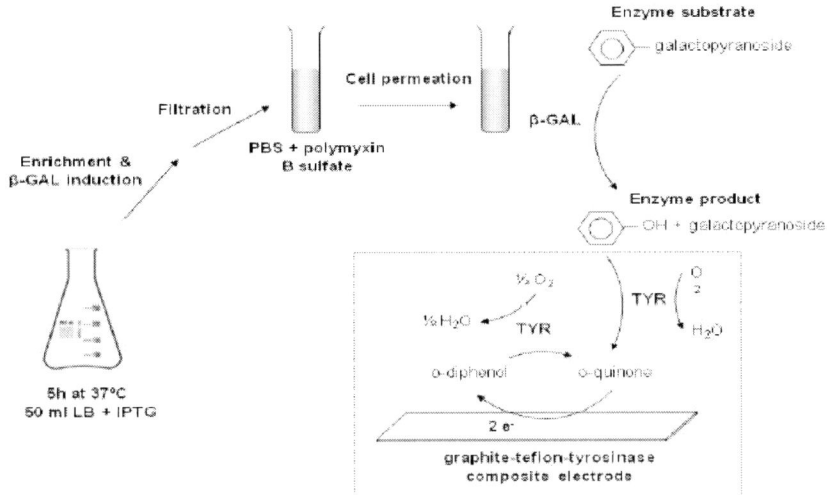

Figure 17. Amperometric biosensor for coliform determination based on β-galactosidase detection. The biosensor described by Serra and co-workers in 2005 consists of sample pre-enrichment in the presence of IPTG (a β-galactosidase inducer), followed by bacteria concentration by filtration, and bacterial wall disruption using a permease. The enzyme released is then detected amperometrically by adding the appropriate enzyme substrate and detecting the electroactive enzyme product using a novel graphite-teflon-tyrosinase composite electrode. An LOD of 1 CFU/100mL was achieved after sample enrichment for 6.5 h.

Another amperometric sensor for monitoring the growth of coliforms, this time in milk samples, was reported by Lee and co-workers [15]. The authors introduced two electrodes into the milk samples under study, which has been inoculated with methylene blue, an electroactive stain component. Current was then registered over time using a homemade potentiostat. In the absence of bacteria, current reached the steady state in about 60 minutes and remained stable for hours. In the presence of certain bacteria, their metabolism resulted in electrochemical reduction of the methylene blue present and significant change in the current registered by the sensor. The time required for this change in current to occur was directly related to the initial bacterial titter. In this way, the sensor showed a linear trend over the range of 10^2–10^8 CFU/mL following incubations of 2-7 hours at 37°C. Both *E. coli* and *Enterobacter aerogenes* were successfully detected within 6-7 hours of incubation, while other microorganisms such as *Bacillus subtilis*, *Citrobacter sp.*, *Klebsiella oxytoca*, *Lactobacillus sp.*, *Staphylococcus aureus*, *Salmonella typhimurium*, and *Saccharomyces sp.*, did not generate significant changes in current within this time frame.

In a different approach, Serra developed an amperometric sensor for the detection of catalase-positive and catalase-negative bacteria [16]. Catalase is an enzyme that is present in catalase-positive bacteria. It catalyses the decomposition of hydrogen peroxide, a highly toxic by-product of many metabolic processes, into water and oxygen. Bacterial catalase can thus be used as an indicator of aerobic and facultative anaerobic microorganism's presence and is useful to monitor their content in, for example, milk, sewage treatment plants or soil samples. On the other hand, the accumulation of hydrogen peroxide in the medium can indicate its production by catalase-negative bacteria. The sensor described in this work was based on the monitoring of hydrogen peroxide consumption or generation at a graphite–Teflon–peroxidase–ferrocene composite electrode. In order to improve sensor reusability and allow detection in complex samples, such as culture media, a nylon membrane was placed onto the electrode surface that prevented its fouling. The authors detected successfully *E. coli* and *Streptococcus pneumoniae* as model microorganism for catalase-positive and catalase-negative bacteria respectively, at titters of 10^5-10^6 CFU/mL, in assays taking 10-15 minutes and not requiring sample pre-concentration or pre-enrichment steps.

4.2. DETECTION OF WHOLE CELLS OR CELL LYSATES

4.2.1. Sandwich-Based Biosensors

Amperometry has also proved useful for the direct detection of pathogen whole cells and cell lysates. In this case, assays derived from the classical sandwich ELISA are regularly exploited, where a first biorecognition element is attached to the electrode surface, while a second enzyme-labelled element serves as a reporter biocomponent. The enzymes most widely used as labels for amperometric detection are horseradish peroxidase (HRP) and alkaline phosphatase (AP), coupled to the use of an appropriate enzyme substrate. This may be a component that is directly converted by the enzyme into an electroactive product, whose oxidation or reduction is subsequently detected at the electrode surface. Alternatively, redox mediators can be used which are not directly an enzyme substrate, but are oxydised/reduced as a result of the enzyme activity. Using reversible redox substrates allows for enhanced electrochemical detection, as the mediator shuttles electrons between the enzyme active site and the electrode surface. In this case, the rate of mediator oxidation/reduction serves to estimate indirectly the concentration of enzyme present.

The immunosensor described by Rao and co-workers was based on a screen printed electrode, which is by definition a disposable and cheap device, and detected the pathogen *Vibrio cholerae* O1 [17]. This screen printed immunosensor worked in a sandwich assay format, using PAb raised against *Vibrio cholerae* whole cell lysates in rabbit and mice, and a reporter anti-mouse Ab conjugated with AP. The substrate for AP was 1-naphthyl phosphate, and the hydrolysis product, 1-naphthol, was amperometrically detected at a potential of +400 mV *vs* Ag/AgCl reference electrode. The lower limit of detection of this biosensor was 10^5 CFU/mL in 55 min, compared to a traditional ELISA that can detect 10^6 CFU/mL in 4 hours. The same authors reported on improved detection limits of this amperometric format directly assayed in environmental samples, after the inclusion of a 6-hour enrichment step [18]. The detection limit was in this case of 8 CFU/mL in ground water and seawater, and 80 CFU/mL in waste waters for *Vibrio cholerae*.

Functionalisation of the electrode surface implies its physical coverage and does significantly affect electron transfer across it. This has moved some authors to propose that target capture onto a surface placed close to the electrode but physically separated from it should provide enhanced detectability, compared to classical biosensing performed directly onto the sensor [19]. Even if not strictly a biosensor, a good example has been provided by Liébana and co-workers, who coupled magnetic immunocapture of bacteria to their amperometric detection [20]. In this case, monoclonal antibodies towards *Salmonella typhimurium* had been incorporated onto the surface of magnetic particles. Bacteria could then be captured using these modified particles, which allowed both separation from the sample matrix and bacteria pre-concentration. For detection, polyclonal antibodies conjugated to HRP and a magnet-modified electrode were used. The authors reported a limit of detection of 5×10^3 CFU/mL for *Salmonella typhimurium* diluted in LB medium, and of 7.5×10^3 CFU/mL when present in milk diluted 1/10 in LB broth. The assay took about 50 minutes and did not require any sample pre-treatment. Sample pre-enrichment for 6-8 hours contributed to additionally improve detectability down to 1 CFU/mL as required by current regulations.

A completely different approach was provided by Ercole and co-workers, who developed a potentiometric sensor for *E. coli* [21]. Target capture was carried out on delimited capture surface in a sandwich format, using for detection an Ab conjugated to the enzyme urease. When urea was added, it was converted by urease into ammonia (NH_3), inducing a change of pH in the medium. Detection took place at a physically separated surface composed by a pH sensitive membrane. This transducer converted the change in pH into a potential shift, which was proportional to the amount of urease and thus target captured. The system allowed detection down to 10 cells per mL of PBS, with an assay time of about 1.5 hours and minimal interference from other potential water polluting bacteria, such as *Pseudomonas marina*, *Sphaerotilus natans* and *Klebsiella oxytoca*, which is phylogenetically related to *E. coli*.

Fluorescence sensors integrate the other transduction strategy which has been extensively exploited for the detection of sandwich-format assays. In this case, detection of captured bacteria is generally attained by using Ab labelled with fluorescent molecules. This is the case of the fiber-optic biosensor described by Kramer et al. [22]. It was based on the commercial evanescent wave biosensor RAPTOR (Research International, Monroe, WA.). The detection procedure giving the best results consisted in a sandwich immunoassay. It used a polyclonal biotinylated antibody to capture *Cryptosporidium* oocists, attached to a streptavidin functionalized waveguide surface, and a cyanine 5-labelled polyclonal antibody as the reporter biocomponent. When the latter had bound, the fluorophore suffered excitation by the evanescent field produced by irradiation of the guide with a 635 nm laser beam. The fluorophore emitted then light above 650 nm, which was recorded by a photodiode. With this technique a limit of detection of 10^5 oocists/mL was achieved.

The example reported by Geng and co-workers [23], was also based on a sandwich capture Ab immobilized onto a polystyrene fiber-optic waveguide by biotin-streptavidin affinity capture and fluorescently-labelled Ab for detection. *E. coli* O157:H7 cells were detected down to 10^3 CFU/mL directly in growth medium, with no interference by non target bacteria including

Salmonella typhimurium, Pseudomonas fluorescens, Shigella flexneri, Serratia marcescens, Yersinia enterocolitica, Lactobacillus plantarum, Escherichia coli ML35, *Bacillus cereus, Enterococcus faecium*, and *L. monocytogenes*, each of them at a concentration of 10^5 CFU/mL. Furthermore, a single *E. coli* O157:H7 CFU/mL, inoculated in ground beef, could be detected by this method after only 4 hours of enrichment. Wei and co-workers reported a similar system, this time using a PAb directly physisorbed onto the fiber optic surface, and fluorescently labelled MAb [24]. The authors claimed that their sensor detected *Yersinia pestis* present at concentrations of 6×10^1–6×10^7 CFU/mL in about 20 min, with little interference from non-target species such as *Y. pseudotuberculosis, Y. enterocolitica, B. anthracis* and *E. coli*. Furthermore, the sensor showed excellent performance on 39 samples, which included 27 tissues of mice infected with *Y. pestis* and 12 tissues of healthy mice as negative control, and identified successfully 92.6% of the infected tissues and all the tissues containing more than 100 CFU/mL of *Y. pestis*.

In a different approach, Zhu and co-workers reported the detection of *E. coli* O157:H7 using the *integrating waveguide biosensor* [25]. The sensing mechanism of the waveguide biosensor was based on a sandwich Ab assay using biotinylated Abs attached to the inner surface of a neutravidin-coated glass capillary, which was used as a waveguide. Detection of target bacteria was performed by incubation of specific Ab conjugated with the Cy5 fluorescent dye, illumination at a 90°C angle of the waveguide and subsequent collection of the emitted fluorescence from the end of the waveguide. The limit of detection was established in 10 cells per capillary (0.075 mL) with an assay time of less than 3 hours. In this work, the authors also demonstrated the ability of the captured *E. coli* cells to grow inside the capillary, what can be very important for the performance of further studies of the captured cells.

As an alternative to fluorophores, enzymes can be used coupled to the use of appropriate substrates which are converted by the enzyme into fluorescent products. Using enzymes allows a significant level of signal amplification and improved limits of detection [26]. Because this type of sensor has been frequently formatted into microarrays for the simultaneous detection of multiple targets, it will be described in more depth in a later section.

4.2.2. One-Step Reagentless Biosensors

Although sandwich-based sensors provide exceptional selectivity and detectability, they involve numerous incubation and washing steps. This translates into long assays, which are also extremely difficult to incorporate to integrated and lab-on-chip devices. For this reason, an important number of works are devoted to the development of faster, one-step, reagentless sensors. This type of sensors theoretically allows minimal sample manipulation and extremely simple assay performance. Sample deposition onto the ready-to-use sensor and subsequent equipment operation should provide straightforward monitoring of the target capture event in real time. Nevertheless, assays fundamented on capture by a single bioreceptor and capture event reportedly offer significantly lower levels of assay specificity and sensitivity. In addition, surface blocking, essential in order to guarantee minimal non-specific adsorption of unwanted components, gains importance in these formats and often affects negatively signal transduction and hence assay limit of detection.

The direct and reagentless detection of bacteria by impedance spectroscopy has proven successful in this context, with reported limits of detection down to 10^4 CFU/mL for *E. coli* and *Salmonella typhimurium* whole cells assayed in PBS and using Ab as the bioreceptor element [27, 28]. In these works, based on the use of microfabricated interdigitated microelectrodes, bacteria captured on top of or between the electrode fingers behave as big insulator elements. Therefore, they induce important changes in the electrochemical behaviour of the electrode-media interface, as well as in the rate of electron transfer across it. Accordingly, the size and geometry of the electrodes used for the measurement have been shown to be determinant for optimal assay performance. For example, when different interdigitated electrodes were used for bacteria impedimetric detection, the best results were obtained for those displaying the smallest features, which were comparable in size to the target analyte [28]. Unexpectedly, the changes in impedance induced by similar concentrations of *E. coli* and *Salmonella* cells differ, presumably due to differences in surface composition and conductance [27, 28]. Attempts by these authors to detect impedimetrically bacteria directly in culture media were unsuccessful or generated significantly higher detection limits, apparently due to the blocking effect by the medium proteins on the sensor surface. Efforts to improve assay performance by carrying out target pre-concentration from the sample matrix or by generating signal enhancement have been made by using Ab-functionalized nanoparticles and magnetic particles [29, 30]. However, none of these works includes appropriate negative controls performed in the presence of non-target bacteria.

Shabani and co-workers [31] exploited a similar approach for the detection of *E. coli* by impedance spectroscopy, although in this case the recognition element used was the bacteriophage T4 instead of antibodies.

Bacteriophages were in this case cross-linked by EDC chemistry onto the surface of screen printed carbon electrodes, where carboxylic groups had been electrochemically generated. Capture and detection of bacteria could then be performed within 40 minutes, when the onset of lysis due to the bacteriophage replication occurred, generating a limit of detection of 10^4 CFU/mL. As in the previous case, the procedure was performed in a saline buffer, which in this case had been supplemented with $MgSO_4$ in order to permit phage infection.

The impedimetric detection of bacteria directly performed in real sample matrices was reported by Pournaras and co-workers [32]. They used gold macroelectrodes, instead of the microelectrodes described in the above works, which they submitted to electropolymerisation of polytyramine. The amino groups incorporated on surface served for the subsequent cross-linking of Ab, followed by blocking with 2% BSA. With this approach, the authors claimed that incubation of the immunosensors directly in culture media or milk is possible, with no significant interference due to non-specific adsorption of the solution components or unrelated microorganisms. In this way, the incubation provides at the same time growth and capture of bacteria present over time, significantly improving detectability. The system showed a limit of detection of 10 CFU/mL for *S. typhimurium* after an incubation time of 3 h. Das and co-workers also reported detection of *Salmonella* in culture medium down to 10^3 CFU/mL, with minimal interference by non-target *E. coli* [33]. In this case, the macroelectrode consisted on a silicon substrate with a 10 μm thick macroporous layer, which showed randomly distributed pores of 1–2 μm diameter.

Impedimetric assays have also proven successful for virus detection. Cho and co-workers produced an impedimetric biosensor for the detection herpes simplex virus (HSV) [34]. The assay consisted in the detection of the cytopathic effect caused by HSV infection on a monolayer of Vero cells, by monitoring the impedimetric changes occurring on the electrodes after infection. The most sensitive/reliable parameter to study was in this example resistance at the cell junctions, which allowed the detection of the cytopathic effect at as low multiplicity of infection as 0.0006 in 124 h. Although HSV is not waterborne, a similar approach could be developed for the detection of water- or foodborne viruses.

Alternatively, bacteria presence can be monitored by following its adsorption and/or growing onto the sensor surface. This is possible because bacteria show very high tendency to adsorb onto a wide variety of surfaces/materials and this deposition contributes to alter the physico-chemical characteristics of the sensing surface. This strategy has been successfully exploited for at least impedance, SPR, QCM and cantilevers. Although completely unspecific, the results reported for QCM allowed monitoring of *E. coli* adsorption and growth in real-time, with a detection limit of 10^2 cells/mL after a 12-hour incubation in either culture medium or milk [35, 36]. The utilisation of microcantilevers coated with a semisolid growth medium, followed by exposition to *E. coli* and sensor washing, allowed subsequent direct monitoring of bacterial growth over the following 5 hours, with no signal change on negative control chips with no bacteria [37, 38].

Although reports exist on the successful detection of bacterial whole cells by SPR, the results suggest that high cell concentrations are needed to obtain detectable signals in this transduction format: the LODs reported are seldom below 10^5 CFU/mL, [39-44]. This is mainly attributed to the fact that the effective penetration depth of the evanescent field in SPR sensors is of approximately 300 nm. Accordingly, and taking into account their shape and size, only a small portion of the captured bacterial cell is in close contact with the sensor surface and thus contributes to SPR signal changes [45]. Additionally, some instruments, such as the Biacore, average the SPR angle measured over an area of approximately 0.25 mm^2 on the sensor surface. Because bacteria are big in size but do not always occupy regular spaces due to, for example, steric hindrance between cells or different physical positioning on surface, this can translate into signal underestimate [39, 40]. Attempts to provide signal amplification by carrying sandwich assay formats using two different Ab have been reported [42, 43].

Vala and co-workers have recently suggested that the use of modified SPR systems, such as the ones based on long-range surface plasmon (LRSP), could generate improved detection of big targets [46]. This assertion relies in the fact that LRSPs have been demonstrated to propagate along thin metal films when these are embedded between two media with similar refractive index. When this is the case, LRSP sensors are characterized by sensing penetration depth that exceeds those of conventional SPR. In their work, the authors use devices that consist of a 25 nm gold film deposited over a 1200 nm Teflon cover, which exhibits a refractive index similar to that of aqueous solutions. The preliminary results show responses 5 times higher for *E. coli* measured at LRSP than those produced at conventional SPR.

Slightly better results were attributed to SPR detection of lysed bacteria. For example, Taylor and co-workers developed an 8-channel SPR equipment and studied four different bacterial species, including *E. coli* O157:H7, *Salmonella choleraesuis*, *Listeria monocytogenes*, and *Campylobacter jejuni*, which had been previously heat-killed and sonicated [47]. Detection included a signal amplification step by incubation with a second Ab, carried immediately after bacteria capture, which also improved assay specificity. LODs ranging 3.4×10^3 to 1.2×10^5 CFU/mL, depending on the microorganism and/or Abs used, were generated. Furthermore, detection was possible not only in PBS, but even in apple juice following pH adjustment from the native pH 3.7 to 7.4. Measurement reliability was additionally improved by incorporating the use of control channels, where no bacteria were injected, which serve to compensate signal changes potentially produced by temperature variation or by Ab non-specific adsorption.

The development of a SPR biosensor for detection of somatic coliphages (i.e. bacteriophages that infect *E. coli*) to be used as indicators of faecal pollution was addressed by García-Aljaro and co-workers [48]. This sensor consisted in the immobilisation of biotinylated bacteria onto avidin-functionalised gold sensors and detection of bacterial lysis caused by bacteriophages. Bacterial lysis correlated with release of intracellular components, which presumably adsorbed onto the sensor surface. This event induced increase in the SPR signal proportional to the amount of phages having infected the biofilm cells. As few as 1 PFU injected into the chamber could be detected after a phage incubation period of 120 min, which equates to an approximate limit of detection of around 10^2 PFU/mL (Figure 18).

An innovative variation of classical SPR provided exceptional results for the detection of *Legionella pneumophila*. This approach was performed by incorporating an optical fiber to the sensing system and using halogen or LED light for detection [49]. The sensing fiber had a polished side, coated with a 37 nm thick gold layer, which was used as an SPR microsensor. The gold surface was functionalised by cross-linking a commercial MAb against *Legionella* to pre-assembled SAMs via EDC/NHS chemistry. According to the preliminary results, the biosensor showed a detection limit of 10^1 CFU/mL and little interference by non-target *E. coli*.

Optical fibers have been extensively exploited in a quite different approach for the optimisation of fluorescent sensors. The example described by Ko and Grant was an ingenious FRET immunosensor [50]. Functionalisation of the fiber surface was performed using protein G labelled with a FRET acceptor fluorophore and an anti-*Salmonella* antibody labelled with a FRET donor fluorophore. Binding of the target pathogen induces a conformational change in the 3D structure of the antibody, which translates in adjustment of the distance between the two fluorophores. The variation in fluorescence emitted was measured at a custom-built benchtop fluorometer. The authors reported limits of detection of 10^3 cells/mL for *Salmonella* in PBS and of 10^5 CFU/g for *S. typhimurium* inoculated in homogenized pork samples. The assay provided results within a 5-min response time and generated not detectable signal for non target *Listeria monocytogenes*.

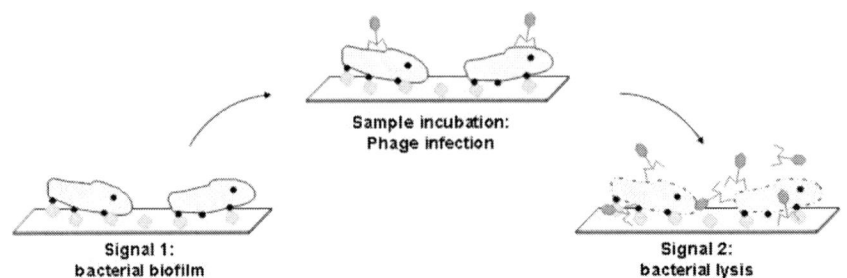

Figure 18. SPR biosensor using bacterial whole cells for bacteriophage detection. In the example described by Garcia-Aljaro in 2008, bacteria are biotinylated and immobilised onto an SPR chip to obtain a biofilm. Incubation in the presence of phages correlates with bacteria infection, which translates into cell disruption and release into the media of important amounts of phage and cell components. The SPR sensor described detects adsorption of such components onto the surface, and thus changes in its optical properties.

As it happens for SPR, few examples of waveguide sensors have been described for bacteria detection. Again, the size of whole bacteria exceeds the evanescent field thickness that determines the extension of the sensitive region of the sensor-solution interface of these devices. In order to improve sensitivity, the penetration depth of the evanescent field should be extended to comprise the entire bacterial cell captured on top of the sensor surface. Zourob and co-workers developed disposable clad leaky waveguide sensors which were produced by simple and inexpensive spin coating techniques [51]. The sensor consisted of a 2 mm thick polymethylmethacrylate chip, which had been subsequently coated with a layer of dye and a 260 nm thick sol–gel layer. According to the authors, the assembly provides an evanescent field penetration depth of about 1.5 μm, compared to the less of 300 nm reported for most related sensors. They detected *Bacillus subtilis* var. *niger* bacterial spores with a detection limit of 8×10^4 spores/mL after a 20-minute incubation. These results were later improved by incorporating an ultrasound pre-concentration step, with reported LOD of 10^3 spores/mL [52].

Piezoelectric biosensors provide another direct reagentless detection strategy. Most of the publications rely on QCM and report limits of detection above 10^5 CFU/mL. This is the case of the work by Si et al, who modified

QCM sensors with monoclonal Ab and detected 10^5-10^8 CFU/mL of *Salmonella enteritidis* within 35 minutes of capture/measurement and LOD of 10^5 CFU/mL. Nearly undetectable signal was generated by similar concentrations of *E. coli* or *Salmonella typhimurium* [53]. In this work, the authors compared the efficiency of several surface immunofunctionalisation strategies over three different surface materials. Su and Li reported an LOD of 10^3 CFU/mL for *E. coli* O157:H7, this time using heat-killed cells [54]. In this work, the sensors had been modified with a long-chain SAM, followed by EDC/NHS conjugation of Ab, blocking with BSA, and immersion into the bacteria-containing sample for 1 hour.

One important limitation of QCM (and some other mechanical sensors) is that the maximal sensitivity is attained when the measurement is performed in the air. Thus, measurement in real time can not be performed. Attempts to optimize QCM measurement in aqueous solution have been described by coupling the sensor to the use of microfluidic channels [55, 56]. This measurement mode, however, often generates lower levels of data reproducibility (*Su and Li 2004*). For example, detection *on-line* has been described for *Edwardsiella tarda*, an economically important Enterobacteriaceae pathogen which is frequently found in organically polluted marine water [57]. In spite of the elevated sensor LOD, 5×10^7 cells/mL, nearly no interference was generated by non-target bacteria. In this work the authors compared three different surface functionalisation strategies and observed better performance of sensors modified with PAb than with MAb. In the example reported by Park and Kim, Abs were thiolated using a cross-linker and self-assembled onto the sensor [58]. *S. typhimurium* was successfully detected in a concentration range of 3.2×10^6–4.8×10^8 CFU/mL in either culture medium and milk solution. Contrary to other works, injection of heat-killed cells generated nearly undetectable signals, similar to those registered for non-target microorganisms. Su and Li reported LODs of 10^5 CFU/mL for heat-killed *Salmonella typhimurium*, either inoculated in PBS or meat extracts [59]. The sensors, which had been modified with protein A by random physisorption, followed by Ab capture, surface blocking, and sample incubation for 1 hour, showed low levels of cross-reactivity towards non-target bacteria such as *E. coli*. Interestingly, sample pre-concentration by magnetic particle immunocapture for 60 additional minutes allowed detection of down to 10^2-10^3 CFU/mL.

Shen et al, on the other hand, proposed that the low levels of detectability reported for big targets on QCM sensors were mainly caused by capture via flexible interactions [60]. They demonstrated that a tighter entrapment, this time using simultaneously a carbohydrate (mannose) on surface and a lectin (concavalin A) in solution as the recognition biocomponents, generated stronger adhesion of *E. coli* cells to the surface, increased contact area between them, more rigid and strong attachment, and improved detectability. Specifically, they obtained an assay linear range between 7.5 x 10^2 and 7.5 x 10^7 cells/mL, with little non-specific adsorption by *Staphylococcus aureus*, compared to LODs above 10^7 CFU/mL for bacteria capture via more flexible capture by just mannose immobilized on surface.

Magnetoelastic biosensors have also been successfully employed for the detection of whole bacterial cells and spores [61, 62]. This kind of sensors are made of a ferromagnetic and amorphous material (usually alloys comprised of iron, nickel, molybdenum and boron). When an oscillating magnetic field is applied, the device undergoes a change in dimensions and vibrates at a certain mechanical resonance frequency, which depends on the material properties as well as the geometry of the sensor. As the target binds to the ligand-modified sensor, there is an increase in the sensor's mass, which in turn causes a corresponding decrease in the sensor's resonance frequency. The mechanical vibrations of a magnetolastic sensor can be detected optically (from the amplitude modulation of a reflected laser beam), acoustically (using a microphone or hydrophone) or using a non-contacting pickup coil to detect the magnetic flux emitted by the sensor. Interestingly, signal recording can be carried out remotely and wirelessly, which allows monitoring in real-time under a variety of conditions. In the example reported by Guntupalli et al, the sensors had been modified with antibodies specific for *Salmonella typhimurium* using the Langmuir–Blodgett monolayer technique. The functionalized sensors were used to study *S. typhimurium* at concentrations 5×10^1 to 5×10^8 CFU/mL, and showed a detection limit of 5×10^3 CFU/mL with no cross-binding to non-target *E. coli* O157:H7 and *Listeria monocytogenes*. The same team applied later this strategy to detection of *Bacillus* spores [61]. In this case, the sensor surface had been functionalized with genetically engineered phages as the capture biocomponent. Concentrations ranging 10^2-10^8 spores/mL were successfully detected, while nearly no signal registered on the negative control sensors with no phages.

Better detection limits have been achieved when using minute-size cantilevers. Campbell and Mutharasan have extensively worked on whole cell bacteria and bacterial spore detection using microfabricated silicon nitride cantilevers. In their work, they have reported successful detection of down to 50-100 CFU/mL of *E. coli* O157 cells, even in complex matrices such as culture media with/without ground meat [63-65], based on detection of the cantilever bending stress. The experiment included sample pre-enrichment for 2-4 hours at 37°C, flowing of 3 mL of pre-enriched sample over the sensor surface, and measurement for about 1 hour. In the same way, these authors, as well as Davila and co-workers, reported detection of tenths of *Bacillus* spores, and even LODs of 5 spores/mL [66]. However, these works have not always been appropriately validated to demonstrate device specificity against non-target bacteria.

Although cantilevers are still a not completely accessible technology, Dhayal and co-workers demonstrated that commercially available cantilever array chips, each of them containing 8 cantilevers, could be modified with gold and specific peptides, and be subsequently used to detect bacterial spores [67]. While their system showed limited sensitivity, with concentrations below 10^5 spore/mL not being detected, significantly lower signals were generated by the negative controls, carried out on sensors modified with a negative control peptide, or in the presence of non-target spores. In addition, the authors demonstrated that experiments performed in static mode produced higher capture efficiencies than experiments where the sample was being flow-pumped. In any case, one of the main limitation of cantilevers is that, due to their reduced size, only samples of small volume can be directly assayed and the analysis of larger samples requires sample pre-treatment and bacteria pre-concentration. Additionally, Ramos and co-workers have demonstrated that the type and magnitude of the signal measured are directly affected by the distribution of the bacterial cells over the surface of the cantilever and not just by the change in mass charge [68]. According to their results, those bacteria captured near the free cantilever end would contribute to a higher extent to the signal recorded when measuring changes in mass charge, while placing bacteria near the clamping region would favour measurement of the variation in flexural rigidity.

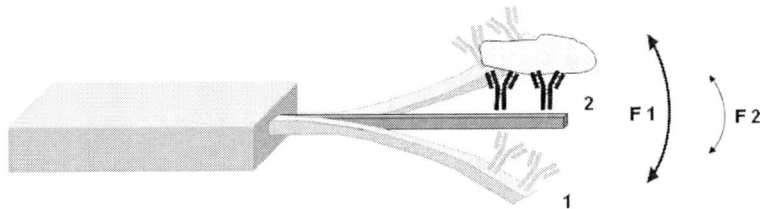

Figure 19. Resonant cantilever for protozoa whole cell detection. In a resonant cantilever sensing format, target capture translates into changes on the sensor resonant frequency due to the change in mass load.

Detection of protozoa with cantilevers was also investigated by Campbell and Mutharasan (Figure 19) [69]. A piezoelectric glass was functionalized with 3-aminopropyl-triethoxysilane and the biorecognition element used, an immunoglobulin M (IgM), was cross-linked via EDC/NHS chemistry. The sensor's resonance frequency showed a logarithmic correlation to concentrations of 100 to 10,000 oocysts/mL, with a detection time of less than 15 min.

Finally, the new materials brought about by the recent advances in nanotechnology have shown excellent properties for the detection of microorganisms. In this way, single-walled carbon nanotubes (SWCN) have been shown to be useful for the detection of *Salmonella infantis* at as low concentrations as 10^2 CFU/mL after 1 hour of exposure using field effect transistors (FET) [70].

4.3. BIOSENSORS FOR THE DETECTION OF BACTERIAL NUCLEIC ACIDS

The efforts made by researchers during the last decade to obtain genetic information from different organisms have generated huge genomic databases where genetic information about a desired gene belonging to a determined species can be found. This information is extremely useful for diagnostic purposes and, together with the advances in the biosensor technology, is the key for the development of simple, fast, cheap, reusable and high-throughput miniaturized and mass-producible analytical devices. Nucleic acid-based biosensors, also known as genosensors, take advantage of the inherent properties of nucleic acids, like their long term stability, and the high selectivity brought about by the detection by hybridisation of complementary nucleic acid probes (or peptide nucleic acids analogues), fixed directly onto the surface of a physical transducer (Figure 20). However, compared to enzyme biosensors and immunosensors, there are still few genosensors in the market. Different transduction systems can be used for the detection of nucleic acids which are next summarized together with some examples of each type.

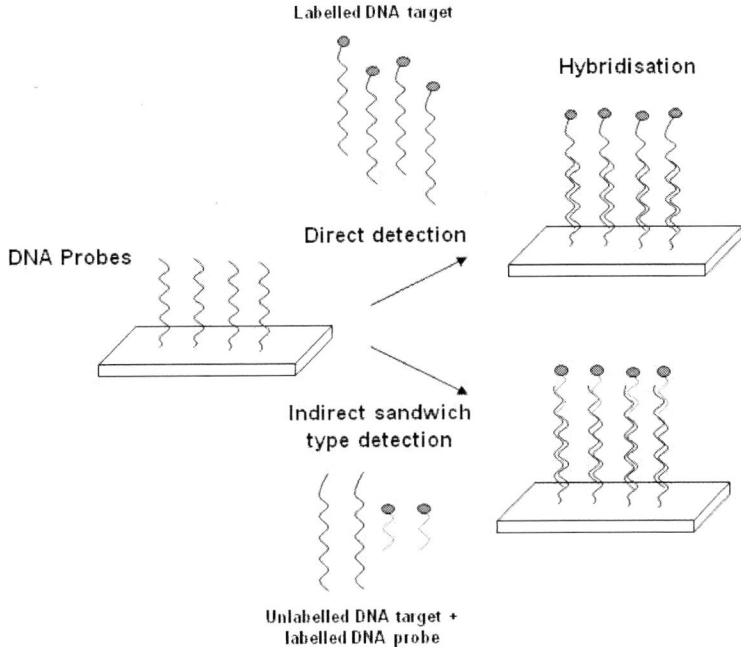

Figure 20. Main formats used for target DNA detection in DNA biosensors. In DNA biosensors detection of target DNA is mainly performed by two different mechanisms: directly after labelling target DNA or in a sandwich format assay using a capture probe and a secondary labelled probe.

4.3.1. Optical Detection

The most common fluorescent genosensors use an optical fiber both as a sensor, thus as physical substrate for bioreceptor immobilisation, and to direct a light beam towards the transducer surface and/or guide back to the detector the fluorescence emitted by sample or assay components. At present, detection is mostly based on the employment of fluorescent labels, such as fluorescein isothiocyanate (FITC), which can be incorporated to DNA probes or directly to the target nucleic acid previous to capture (Figure 20). Nevertheless, detection of the hybridisation event was initially carried out using ethidium bromide, which intercalates non-specifically between the base pairs and in the major grooves of double stranded DNA (dsDNA) [71]. More recently, quantum dots have appeared as an alternative to conventional fluorophores.

Quantum dots are semiconductor particles (e.g. ZnS and CdSe) that can be used as labels in the same way as classical fluorophores, but that provide superior performance and assay sensitivity because of their higher fluorescence emission and photostability [72]. Quantum dots are used in assay formats similar to those previously reported. For instance, detection of *Bacillus anthracis* in a microarray format using a fiber optic biosensor was described by Shepard. In this work, beads coated with ssDNA probes were immobilised in the wells of a fiber-optic array. Samples containing bacteria were then amplified by PCR using biotinylated primers and submitted to hybridisation onto the bead-fiber array. Captured sequences were subsequently labelled by incubation with streptavidin-quantum dot conjugates [73]. Although a single microorganism was detected, simultaneous monitoring with eight different reporters was performed in parallel, demonstrating the methodology potential for multiplexed analysis.

Molecular beacons have also attracted the interest of researchers to be used as alternative probes for the detection of nucleic acids, mainly because they involve a completely reagentless assay format. Molecular beacons are ssDNA probes specifically designed to form a stem-and-loop structure, where the loop portion is complementary to the target sequence, flanked by two small segments complementary to each other (Figure 21). A fluorophore and a quencher are situated at the ends of the stem so that the fluorescence of the fluorophore is quenched by energy transfer when the molecular beacon is in the hairpin closed shape. However, in the presence of the target sequence, the hairpin structure opens because the probe/target hybrid is more thermodynamically stable. In consequence, the fluorophore and the quencher are separated, allowing the probe to emit a fluorescence signal [74]. This kind of probes were used by Kong and co-workers to construct a biosensor for the detection of *E. coli*, *S. enterica* and *S. dysenteriae* in seawater samples. Three genes targeting for the different pathogens were PCR-amplified and then detected using a molecular beacon. The assay showed a limit of detection of 10^1 CFU/mL [75]. A slightly higher limit of detection was reported by Fortin and co-workers for detection of *E. coli* O157:H7 in raw milk and apple cider: 10^2 CFU/mL [76]. The detection limit was subsequently improved by two orders of magnitude to 1 CFU/mL after inclusion of a six-hour enrichment.

The development of new SPR-based systems employing polarization control and internal referencing has provided new analytical tools with better limits of detection. This has been profited by Piliarik and co-workers to develop a SPR-based DNA array for the detection of foodborne bacterial nucleic acids [77]. The target bacteria, *Brucella abortus*, *E. coli* and

Staphylococcus aureus, were detected at a concentration of 100 pM in less than 15 min without the need of any label.

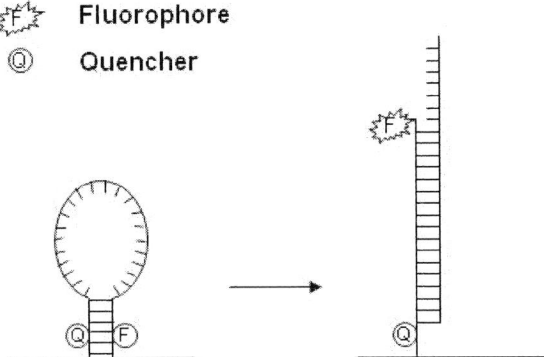

Figure 21. Molecular beacons. In the absence of target, the molecular beacon probe forms a stem-and-loop structure, where the fluorophore is located close to the quencher eliminating the ability of the fluorophore to fluoresce. In the presence of target, the target-probe hybrid is more stable than the stem-and-loop structure, and the molecular beacon undergoes a conformational change that forces the fluorophore away from the quencher.

One of the main limitations of most DNA sensors is the lack of discrimination between viable and non-viable cells. In this context, Baeumner and co-workers developed a biosensor specific for viable *E. coli* cells based on the detection of the mRNA codifying for a heat shock protein (*clpB*) [78]. The mRNA was extracted from the cells after a heat shock treatment, purified and amplified using the isothermal nucleic acid sequence-based amplification technique. The amplified mRNA was detected in only 15-20 min after signal amplification using tagged liposomes encapsulating sulphorodamine B fluorophore. The reflectance of the samples was measured with a reflectometer at a wave length of 560 nm. Although a good limit of detection was claimed by the authors (40 CFU/mL or 5 fmol of target sequence), again the effect of different matrices on the assay was not determined.

4.3.2. Electrochemical Detection

Farabullini and co-workers developed an electrochemical genosensor for the simultaneous detection of *Salmonella enterica*, *Listeria monocytogenes*, *Staphylococcus aureus*, and *E. coli* O157:H7 [79]. The assay involved first

PCR amplification of specific target genes for each bacteria, followed by a DNA sandwich hybridisation assay using biotinylated probes. The resulting biotinylated hybrids were coupled to streptavidin-AP and electrochemically detected with α-naphthyl phosphate using differential pulse voltammetry. The hybridisation assay together with the electrochemical detection of the target sequence could be performed in 1 h, with sensitivity at the nanomolar level. In a similar way, a DNA biosensor for the detection of human cytomegalovirus based on the electrochemical detection of amplified viral sequences was developed [80]. The amplimers were immobilised onto the surface of the sensor and hybridised with biotinylated probes. However, in this case biotin was detected using streptavidin conjugated to horseradish peroxidase enzyme and measuring by differential pulse voltammetry the electroactive product 2,2'-diaminobenzene enzymatically generated from the *o*-phenylendiamine substrate . The sensor showed a limit of detection of 0.6 amol/mL.

In order to shorten the time of analysis, Elsholz and co-workers developed a sensor that targeted the 16S rRNA. The assay took profit of the fact that bacteria carry multiple copies of the gene coding for this RNA, which makes PCR amplification unnecessary [81]. This gene also carries signature sequences that are species specific and is then a good candidate to develop probes for detection of pathogenic bacteria. The developed biosensor consisted in an electrical microarray for the simultaneous detection of *E. coli*, *P. aeruginosa*, *Enterococcus faecalis*, *S. aureus* and *S. epidermidis*. Thiol-modified oligonucleotides used as capture probes where immobilised by self-assembly onto gold interdigitated array electrodes. Detection of target RNA was performed in a sandwich format by hybridising biotinylated oligonucleotide probes. The assembly was then amperometrically detected by redox cycling after incubation with avidin-AP, which transforms the substrate *p*-aminophenyl phosphate into the electroactive molecule *p*-aminophenol. The detection time for this reaction was of around 15 min and in the case of *E. coli* the limit of detection was established in 0.5 ng/µl (approximately 10^5 CFU/mL).

The use of magnetic nanoparticles coated with DNA probes for detection of the target microorganisms has been shown useful for the isolation and concentration of DNA targets from the sample previous to PCR. Taking advantage of this concept, Loaiza and co-workers developed a highly sensitive genosensor for the detection of coliforms based on the detection of the *lacZ* gene [82]. This approach exploited biotinylated capture probes attached to streptavidin-modified magnetic beads, which were used to capture the product generated by asymmetric PCR amplification of target DNA with biotinylated primers. Detection was performed amperometrically in the presence of streptavidin-HRP conjugates that mediated the reduction of H_2O_2 in the presence of tetrathiafulvalene-modified gold-screen printed electrodes. The assay was performed on *E. coli* as a model microorganism, with successful detection of titters as low as 0.01 CFU/mL.

In an attempt to produce a self-contained and fully integrated biochip for the detection of pathogens, Liu and co-workers developed a device that conducted from the processing of the sample to the final detection step [83]. This was achieved by a series of electromechanical valves and thermopneumatic pumps, with no external operations required, diminishing the possibility of contamination of the samples. The whole procedure involved immunomagnetic separation, followed by PCR amplification and electrochemical detection. In spite of the high limit of detection (10^6 *E. coli* in 1 mL of a complex matrix, equivalent to 5 ng of DNA), detectability could be improved by pre-enrichment and/or pre-concentration of the sample.

Combination of quantitative PCR with electrochemical HRP-mediated detection has also generated very sensitive biosensors. For example, a limit of detection of 0.45 ng/µL was obtained after only 10 cycles of amplification for the detection of pathogenic *E. coli* using the *eaeA* gene as target [84].

A different approach for the detection of *E. coli* O157 was developed by Liao and co-workers [85]. The approach was based on a competitive assay using reporter DNA-tagged liposomes containing the electroactive redox marker $Ru(NH3)_6^{3+}$. It targeted detection of the *rfbE* gene, a gene involved in the synthesis of the O-antigen which is highly conserved among the O157 serotype, onto screen printed electrodes using square wave voltammetry. The authors demonstrated detection of the *rfbE* DNA at a concentration of as low as 0.75 amol based on detection of the released $Ru(NH3)_6^{3+}$. A schematic representation of the assay is shown in Figure 22.

Finally, Wang and co-workers demonstrated detection of *E. coli* O157:H7 without the need of any label in a PCR extension hybridisation assay using impedance and cyclic voltammetry to measure changes in conductivity [86]. The assay was based on the immobilisation of single stranded DNA probes onto aluminium anodized oxide nanopore membranes, followed by hybridisation of target DNA and a polymerase mediated extension of the immobilised probes using the target DNA as template. The limit of detection of this assay was of 500 pmol of DNA.

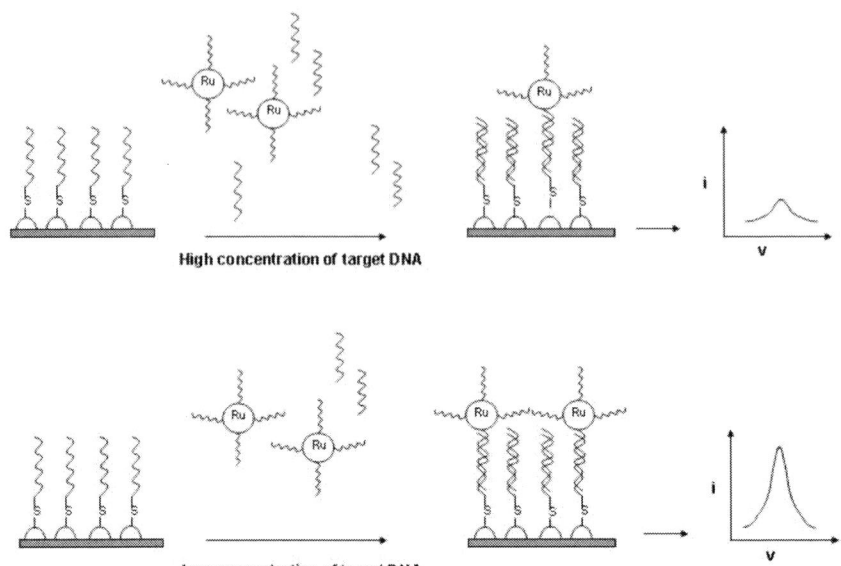

Figure 22. Competitive assay using reporter DNA-tagged liposomes Ru(NH$_3$)$_6^{3+}$. In the assay reported by Liao and Ho, detection of the *rfbE* gene from *E. coli* O157 was performed in a competitive assay format using capture DNA probes and DNA-tagged liposomes containing Ru(NH$_3$)$_6^{3+}$ to compete with target DNA in an hybridisation assay. In consequence, the increase in current upon hybridisation is inversely proportional to the target DNA concentration, which is directly proportional to the amount of DNA-tagged liposomes hybridised with the capture probes.

4.3.3. Piezoelectric Genosensors

The piezoelectric genosensors reported are mainly based on the utilisation of QCM. In this respect, QCM has generated significantly better LODs for detection of bacterial nucleic acids than for detection of whole cells and other cell components. Mao and co-workers developed a QCM DNA sensor for the detection of *E. coli* O157:H7, using the nanoparticle amplification method to ameliorate the limit of detection of the system [87]. The target gene was again *eaeA*, which codes for intimin. Detection was carried out by asymmetric PCR amplification (a technique that predominantly produces ssDNA) using biotinylated primers, followed by hybridisation with thiolated DNA probes previously immobilized onto the surface of the QCM sensor. The signal was subsequently amplified by attachment of streptavidin-conjugated Fe$_3$O$_4$ nanoparticles to the biotinylated PCR products, which produced additional increase in the mass load and resulted in increased frequency change. The limit of detection was around 10^2 CFU/mL and the detection of the amplified PCR products could be performed within 30 min. An alternative amplification methodology includes the use of a second thiolated probe conjugated to gold nanoparticles to recognise the hybridised target sequences [88]. This method generated similar detection limits than the one described above and was achieved in a total assay time, including processing of the samples, of 3h. Apparently, the use of sample pre-enrichment was unnecessary, even when the assay was directly performed in food samples. Inclusion of a short pre-enrichment step could presumably lower the detection limit down to 1 CFU/mL.

A label free QCM based biosensor for the detection of *E. coli* O157:H7 was developed by Wu and co-workers [89]. The assay involved first PCR amplification of a specific gene coding for a virulence factor, the intimin (*eaeA*), followed by hybridisation of the target DNA onto the gold surface of the QCM using a circulating-flow system and without the use of any label for the detection. The system was able to detect amplified DNA from real samples and can be applied to detection of other pathogens as long as specific primers and probes are used. In a similar approach, Sun and co-workers were able to improve the limit of detection provided by other biosensors down to 23 *E. coli* by adjusting the hybridisation parameters in a flow system to avoid renaturalization of DNA before hybridisation with the probe [90].

4.4. DETECTION OF TOXINS

Some authors have addressed the detection of bacterial toxins pathogenic to humans, such as the cholerae toxin (CT). For example, Rowe-Taitt and co-workers were able to detect a CT concentration of as low as 1.6 ng/mL with an evanescent wave fluorescence biosensor [91], while Zayats and co-workers reported a limit of detection of 10^{-11} M by impedance measurement after capture of CT onto the antibody modified surface of an ISFET biosensor [92]. O'Brien, on the other hand, exploited a previously developed bidiffractive grating biosensor for the simultaneous immunodetection of four different targets [93]. Two bacterial toxins were included among them: *Staphylococcus aureus* enterotoxin B (SEB) and *Clostridium botulinum* toxin (BOT). Target binding induces in this case changes in the refractive index measured on the surface of a plastic waveguide over time. However, and in spite of the minucious optimisation of the device performance, the work lacks the appropriate negative controls.

4.5. MULTIPLEX DETECTION

The simultaneous detection of different microorganisms is advantageous as it reduces, among others, the time necessary to perform sample analysis, as well as the time and level of sample manipulation by the workers. The most frequently used formats for multiplex detection make use of the microarray technology, and predominantly rely on detection of DNA coding for virulence markers or 16S rRNA signature sequences of the different target pathogens by using oligonucleotide probes as the capture element. One of the first microarrays described for the detection of pathogens was developed by Wilson and co-workers [94]. This assay targeted the simultaneous detection of 11 bacterial, five viral, and two eukaryotic pathogens. In order to diminish the rate of false positives, the authors used multiple probes for each pathogen, what ensured positive results for any target present at a relative abundance between 1 and 5%. Miller et al. have recently exploited a similar approach for the simultaneous detection of 12 different bacteria in spiked water samples (*Aeromonas hydrophila*, *Helicobacter pylori*, *Legionella pneumophila*, *Pseudomonas aeruginosa*, *Vibrio cholerae*, *Vibrio parahaemolyticus*, *Yersinia enterocolitica*, *Clostridium perfringens*, *Salmonella*, *Staphylococcus aureus*, *Campylobacter jejuni*, and *Listeria monocytogenes*) [95]. Although the limits of detection were higher than in the previous example, detection was successfully improved by incorporating a multiplex PCR step before the microarray detection. In this way, positives were detected when present at a relative abundance of 0.1 and 0.01%, depending on the pathogen and the target gene.

Bavykin and co-workers developed a system for identification of bacteria based on the hybridisation of 16S rRNA fragments fluorescently labelled (obtained after lysis of whole cells) with 16S DNA probes immobilised onto a glass slide [96]. This approach was designed for the detection of log phase cultures of *E. coli*, *B. subtilis* and *B thuringensis* in a total processing time of 50 min, even though a first enrichment of the samples was necessary in case of samples containing low numbers of pathogens.

In a different approach, LaGier and co-workers designed a multiplex assay for the detection of 8 water contaminants (the red dinoflagellate *Karenia brevis*, the faecal indicator bacteria *Enterococcus* spp., the markers of faecal pollution of human source (*Bacteroides* HF8 specific-human cluster, and the *esp* gene of *Enterococcus faecium*), the bacterial pathogens *Escherichia coli* 0157:H7, *Salmonella* spp., *Campylobacter jejuni,* and *Staphylococcus aureus*, and the viral pathogen adenovirus) [97]. This detection format combined a PCR assay, conducted for each microorganism using a specific biotinylated primer in order to produce biotinylated amplicons, and electrochemical detection of the generated amplicons, in less than 5 hours. Briefly, the biotinylated PCR products were immobilised onto a neutravidin coated 8-well SPE sensor strip and detected via horseradish peroxidase chemistry by intermittent pulse amperometry. The amount of DNA bound to the sensor was proportional to the electrochemical current generated in the presence of the enzyme substrate during the application of milliseconds pulses of -100 mV. In contrast to the majority of the assays reported, this assay was performed on natural coast waters and sediments, showing its potential applicability to real samples. Disappointingly, the sensitivity for each microorganism was not strictly determined.

Sapsford *et al* developed a versatile fluorescence immunoarray for the detection of foodborne contaminants [98]. Glass slides were silanised, conjugated to neutravidin, and different biotinylated Ab were patterned using a 6- or a 12-channel flow cell for their directed affinity capture. The combination of sandwich and competition assay formats allowed the authors to detect both big and small targets, including whole *Campylobacter jejuni* cells and the mycotoxins ochratoxin A, fumonisin B, aflatoxin B1 and deoxynivalenol. The LODs reported ranged from 5×10^2 to 4×10^3 CFU/mL for *C. jejuni* present in different food matrices such as yogurt, milk, ground turkey sausage and ground turkey ham, and down to 0.3 ng/mL for toxins in an assay of less than half an hour.

Multiplex detection optimisation has also benefited from the recent advances in SPR and SPR imaging technology, which allow carrying multi-channel measurements. For example, a multiplex detection format was developed for the detection of *Escherichia coli* O157:H7, *Salmonella choleraesuis* serotype typhimurium, *Listeria monocytogenes*, and *Campylobacter jejuni* using an 8-channel SPR sensor [47]. The bacteria where first heat killed and sonicated in order to improve the limit of detection, which was established in a range from to 3.4×10^3 to 1.2×10^5 CFU/mL for the different species, either alone or in a mixture comprising the four of them in PBS pH 7.4 (Figure 23).

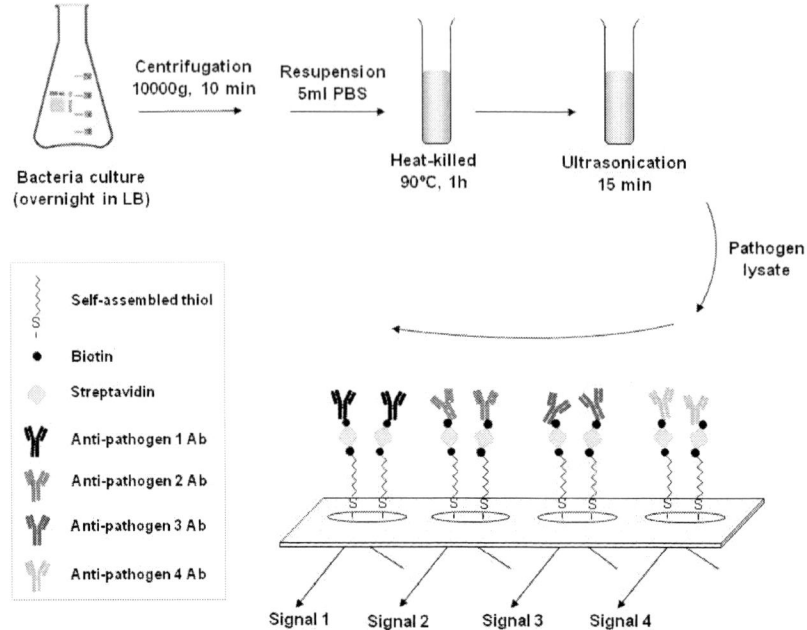

Figure 23. Bacteria detection using multiplex SPR. In spite of its limited performance for the recognition of whole cells, SPR has shown applicable for the detection of bacteria lysates. Coupled to the most recent advances in SPR technology, multiplexed detection is possible by studying in parallel sensors modified with different capture biocomponents. In the example described by Taylor and co-workers, bacteria were lysed by heat inactivation and sonication previous to their analysis.

REFERENCES

[1] E. Baldrich, N. Vigues, J. Mas and F. X. Munoz (2008). Sensing bacteria but treating them well: Determination of optimal incubation and storage conditions. *Analytical Biochemistry, 383*, 68-75.

[2] M. W. Griffiths (1996). The role of ATP bioluminescence in the food industry: New light on old problems. *Food Technology, 50*, 62-72.

[3] K. O. Colquhoun, S. Timms and C. R. Fricker (1998). A simple method for the comparison of commercially available ATP hygiene-monitoring systems. *Journal of Food Protection, 61*, 499-501.

[4] J. Luo, X. Liu, Q. Tian, W. Yue, J. Zeng, G. Chen and X. Cai (2009). Disposable bioluminescence-based biosensor for detection of bacterial count in food. *Analytical Biochemistry, 394*, 1-6.

[5] P. Cady (1975). Rapid automated bacterial identification by impedance measurement. In C.-G. Heden and T. Illeni, *New Approaches to the Identification of Microorganisms* (73-99).New York, John Wiley

[6] A. Ur and D. Brown (1975). Impedance monitoring of bacterial activity. *Journal of Medical Microbiology*, *8*, 19-28.

[7] P. A. Noble, M. Dziuba, D. J. Harrison and W. L. Albritton (1999). Factors influencing capacitance-based monitoring of microbial growth. *Journal of Microbiological Methods*, *37*, 51-64.

[8] L. J. Yang and Y. B. Li (2006). Detection of viable *Salmonella* using microelectrode-based capacitance measurement coupled with immunomagnetic separation. *Journal of Microbiological Methods*, *64*, 9-16.

[9] M. D. Morales, B. Serra, A. Guzman-Vazquez de Prada, A. J. Reviejo and J. M. Pingarron (2007). An electrochemical method for simultaneous detection and identification of *Escherichia coli*, *Staphylococcus aureus* and *Salmonella choleraesuis* using a glucose oxidase-peroxidase composite biosensor. *Analyst*, *132*, 572-578.

[10] B. Serra, M. D. Morales, J. Zhang, A. J. Reviejo, E. H. Hall and J. M. Pingarron (2005). In-a-day electrochemical detection of coliforms in drinking water using a tyrosinase composite biosensor. *Analytical Chemistry*, *77*, 8115-8121.

[11] Y. Cheng, Y. Liu, J. Huang, K. Li, Y. Xian, W. Zhang and L. Jin (2009). Amperometric tyrosinase biosensor based on Fe_3O_4 nanoparticles-coated carbon nanotubes nanocomposite for rapid detection of coliforms. *Electrochimica Acta*, *54*, 2588-2594.

[12] I. H. Boyaci, Z. P. Aguilar, M. Hossain, H. B. Halsall, C. J. Seliskar and W. R. Heineman (2005). Amperometric determination of live *Escherichia coli* using antibody-coated paramagnetic beads. *Analytical and Bioanalytical Chemistry*, *382*, 1234-1241.

[13] T. Neufeld, A. Schwartz-Mittelmann, D. Biran, E. Z. Ron and J. Rishpon (2003). Combined phage typing and amperometric detection of released enzymatic activity for the specific identification and quantification of bacteria. *Analytical Chemistry*, *75*, 580-585.

[14] M. Yemini, Y. Levi, E. Yagil and J. Rishpon (2007). Specific electrochemical phage sensing for *Bacillus cereus* and *Mycobacterium smegmatis*. *Bioelectrochemistry*, *70*, 180-184.

[15] Y. G. Lee, H. Y. Wu, C. L. Hsu, H. J. Liang, C. J. Yuan and H. D. Jang (2009). A rapid and selective method for monitoring the growth of coliforms in milk using the combination of amperometric sensor and

reducing of methylene blue. *Sensors and Actuators. B: Chemical*, *141*, 575-580.

[16] B. Serra, J. Zhang, M. D. Morales, A. G.-V. de Prada, A. J. Reviejo and J. M. Pingarrón (2008). A rapid method for detection of catalase-positive and catalase-negative bacteria based on monitoring of hydrogen peroxide evolution at a composite peroxidase biosensor. *Talanta*, *75*, 1134-1139.

[17] V. K. Rao, M. K. Sharma, A. K. Goel, L. Singh and K. Sekhar (2006). Amperometric immunosensor for the detection of *Vibrio cholerae* O1 using disposable screen-printed electrodes. *Analytical Sciences*, *22*, 1207-1121.

[18] M. K. Sharma, A. K. Goel, L. Singh and V. K. Rao (2006). Immunological biosensor for detection of Vibrio cholera O1 in environmental water samples. *World Journal of Microbiology and Biotechnology*, *22*, 1155-1159.

[19] E. Baldrich, F. J. del Campo and F. X. Muñoz (2009). Biosensing at disk microelectrode arrays. Inter-electrode functionalisation allows formatting into miniaturised sensing platforms of enhanced sensitivity. *Biosensors and Bioelectronics*, *25*, 920-926.

[20] S. Liebana, A. Lermo, S. Campoy, M. P. Cortes, S. Alegret and M. I. Pividori (2009). Rapid detection of *Salmonella* in milk by electrochemical magneto-immunosensing. *Biosensors and Bioelectronics*, *25*, 510-513.

[21] C. Ercole, M. del Gallo, M. Pantalone, S. Santucci, L. Mosiello, C. Laconi and A. Lepidi (2002). A biosensor for *Escherichia coli* based on a potentiometric alternating biosensing (PAB) transducer. *Sensors and Actuators, B: Chemical*, *83*, 48-52.

[22] M. F. Kramer, G. Vesey, N. L. Look, B. R. Herbert, J. M. Simpson-Stroot and D. V. Lim (2007). Development of a *Cryptosporidium* oocyst assay using an automated fiber optic-based biosensor. *Journal of Biological Engineering*, *1*, art. no. 3.

[23] T. Geng, J. Uknalis, S.-I. Tu and A. K. Bhunia (2006). Fiber-optic biosensor employing alexa-fluor conjugated antibody for detection of *Escherichia coli* O157:H7 from ground beef in four hours. *Sensors*, *6*, 796-807.

[24] H. Wei, Y. Zhao, Y. Bi, H. Liu, Z. Guo, Y. Song, J. Zhai, H. Huang and R. Yang (2007). Direct detection of *Yersinia pestis* from the infected animal specimens by a fiber optic biosensor. *Sensors and Actuators, B: Chemical*, *123*, 204-210.

[25] P. Zhu, D. R. Shelton, J. S. Karns, A. Sundaram, S. Li, P. Amstutz and C. M. Tang (2005). Detection of water-borne *E. coli* O157 using the integrating waveguide biosensor. *Biosensors and Bioelectronics, 21,* 678-683.

[26] J. M. Song and T. Vo-Dinh (2004). Miniature biochip system for detection of *Escherichia coli* O157 : H7 based on antibody-immobilized capillary reactors and enzyme-linked immunosorbent assay. *Analytica Chimica Acta, 507,* 115-121.

[27] S. M. Radke and E. C. Alocilja (2005). A microfabricated biosensor for detecting foodborne bioterrorism agents. *IEEE Sensors Journal, 5,* 744-750.

[28] O. Laczka, E. Baldrich, F. X. Munoz and F. J. del Campo (2008). Detection of *Escherichia coli* and *Salmonella typhimurium* using interdigitated microelectrode capacitive immunosensors: The importance of transducer geometry. *Analytical Chemistry, 80,* 7239-7247.

[29] M. Varshney, Y. Li, B. Srinivasan and S. Tung (2007). A label-free, microfluidics and interdigitated array microelectrode-based impedance biosensor in combination with nanoparticles immunoseparation for detection of *Escherichia coli* O157:H7 in food samples. *Sensors and Actuators, B: Chemical, 128,* 99-107.

[30] G. Kim, J. H. Mun and A. S. Om (2007). Nano-particle enhanced impedimetric biosensor for detedtion of foodborne pathogens. *Journal of Physics: Conference Series, 61,* 555-559.

[31] A. Shabani, M. Zourob, B. Allain, C. A. Marquette, M. F. Lawrence and R. Mandeville (2008). Bacteriophage-Modified Microarrays for the Direct Impedimetric Detection of Bacteria. *Analytical Chemistry, 80,* 9475-9482.

[32] A. V. Pournaras, T. Koraki and M. I. Prodromidis (2008). Development of an impedimetric immunosensor based on electropolymerized polytyramine films for the direct detection of *Salmonella typhimurium* in pure cultures of type strains and inoculated real samples. *Analytica Chimica Acta, 624,* 301-307.

[33] R. D. Das, C. RoyChaudhuri, S. Maji, S. Das and H. Saha (2009). Macroporous silicon based simple and efficient trapping platform for electrical detection of *Salmonella typhimurium* pathogens. *Biosensors and Bioelectronics, 24,* 3215-3222.

[34] S. Cho, S. Becker, T. von Briesen and H. Thielecke (2007). Impedance monitoring of herpes simplex virus-induced cytopathic effect in Vero cells. *Sensors and Actuators, B: Chemical, 123* 978-982.

[35] K. S. Chang, H. D. Jang, C. F. Lee, Y. G. Lee, C. J. Yuan and S. H. Lee (2006). Series quartz crystal sensor for remote bacteria population monitoring in raw milk via the Internet. *Biosensors and Bioelectronics, 21*, 1581-1590.

[36] H.-C. Han, Y.-R. Chang, W.-L. Hsu and C.-Y. Chen (2009). Application of parylene-coated quartz crystal microbalance for on-line real-time detection of microbial populations. *Biosensors and Bioelectronics, 24*, 1543-1549.

[37] A. J. Detzel, G. A. Campbell and R. Mutharasan (2006). Rapid assessment of *Escherichia coli* by growth rate on piezoelectric-excited millimeter-sized cantilever (PEMC) sensors. *Sensors and Actuators, B: Chemical, 117*, 58-64.

[38] K. Y. Gfeller, N. Nugaeva and M. Hegner (2005). Micromechanical oscillators as rapid biosensor for the detection of active growth of *Escherichia coli. Biosensors and Bioelectronics, 21*, 528-533.

[39] P. M. Fratamico, T. P. Strobaugh, M. B. Medina and A. G. Gehring (1998). Detection of *Escherichia coli* O157 : H7 using a surface plasmon resonance biosensor. *Biotechnology Techniques, 12*, 571-576.

[40] V. Koubova, E. Brynda, L. Karasova, J. Skvor, J. Homola, J. Dostalek, P. Tobiska and J. Rosicky (2001). Detection of foodborne pathogens using surface plasmon resonance biosensors. *Sensors and Actuators, B: Chemical, 74*, 100-105.

[41] V. Nanduri, A. K. Bhunia, S.-I. Tu, G. C. Paoli and J. D. Brewster (2007). SPR biosensor for the detection of L. monocytogenes using phage-displayed antibody. *Biosensors and Bioelectronics, 23*, 248-252.

[42] A. Subramanian, J. Irudayaraj and T. Ryan (2006). A mixed self-assembled monolayer-based surface plasmon immunosensor for detection of *E-coli* O157 : H7. *Biosensors and Bioelectronics, 21*, 998-1006.

[43] G. Bokken, R. J. Corbee, F. van Knapen and A. A. Bergwerff (2003). Immunochemical detection of *Salmonella* group B, D and E using an optical surface plasmon resonance biosensor. *Fems Microbiology Letters, 222*, 75-82.

[44] J. Y. Jyoung, S. H. Hong, W. Lee and J. W. Choi (2006). Immunosensor for the detection of *Vibrio cholerae* O1 using surface plasmon resonance. *Biosensors and Bioelectronics, 21*, 2315-2319.

[45] P. Leonard, S. Hearty, J. Quinn and R. O'Kennedy (2004). A generic approach for the detection of whole *Listeria monocytogenes* cells in

contaminated samples using surface plasmon resonance. . *Biosensors and Bioelectronics*, *19*, 1331-1335.

[46] M. Vala, S. Etheridge, J. A. Roach and J. Homola (2009). Long-range surface plasmons for sensitive detection of bacterial analytes. *Sensors and Actuators, B: Chemical*, *139*, 59-63.

[47] A. D. Taylor, J. Ladd, Q. M. Yu, S. F. Chen, J. Homola and S. Y. Jiang (2006). Quantitative and simultaneous detection of four foodborne bacterial pathogens with a multi-channel SPR sensor. *Biosensors and Bioelectronics*, *22*, 752-758.

[48] C. Garcia-Aljaro, X. Munoz-Berbel, A. T. Jenkins, A. R. Blanch and F. X. Munoz (2008). Surface plasmon resonance assay for real-time monitoring of somatic coliphages in wastewaters. *Applied and Environmental Microbiology*, *74*, 4054-4058.

[49] H.-Y. Lin, Y.-C. Tsao, W.-H. Tsai, Y.-W. Yang, T.-R. Yan and B.-C. Sheu (2007). Development and application of side-polished fiber immunosensor based on surface plasmon resonance for the detection of *Legionella pneumophila* with halogens light and 850 nm-LED. *Sensors and actuators. A, Physical 138*, 7.

[50] S. H. Ko and S. A. Grant (2006). A novel FRET-based optical fiber biosensor for rapid detection of *Salmonella typhimurium*. *Biosensors and Bioelectronics*, *21*, 1283-1290.

[51] M. Zourob, S. Mohr, B. J. T. Brown, P. R. Fielden, M. B. McDonnell and N. J. Goddard (2005). Bacteria detection using disposable optical leaky waveguide sensors. *Biosensors and Bioelectronics*, *21*, 293-302.

[52] M. Zourob, J. J. Hawkes, W. T. Coakley, B. J. T. Brown, P. R. Fielden, M. B. McDonnell and N. J. Goddard (2005). Optical leaky waveguide sensor for detection of bacteria with ultrasound attractor force. *Analytical Chemistry*, *77*, 6163-6168.

[53] S. H. Si, X. Li, Y. S. Fung and D. R. Zhu (2001). Rapid detection of *Salmonella enteritidis* by piezoelectric immunosensor. *Microchemical Journal*, *68*, 21-27.

[54] X. L. Su and Y. B. Li (2004). A self-assembled monolayer-based piezoelectric immunosensor for rapid detection of *Escherichia coli* O157: H7. *Biosensors and Bioelectronics*, *19*, 563-574.

[55] B. Godber, M. Frogley, M. Rehak, A. Sleptsov, K. S. J. Thompson, Y. Uludag and M. A. Cooper (2007). Profiling of molecular interactions in real time using acoustic detection. *Biosensors and Bioelectronics 22*, 2382-2386.

[56] B. Godber, K. S. J. Thompson, M. Rehak, Y. Uludag, S. Kelling, A. Sleptsov, M. Frogley and M. A. Cooper (2005). Direct quantification of analyte concentration by resonant acoustic profiling. *Clinical Chemistry 51*, 1962-1972.

[57] S.-R. Hong, S.-J. Choi, H. D. Jeong and S. Hong (2009). Development of QCM biosensor to detect a marine derived pathogenic bacteria Edwardsiella tarda using a novel immobilisation method. *Biosensors and Bioelectronics*, *24*, 1635-1640.

[58] I. S. Park, W. Y. Kim and N. Kim (2000). Operational characteristics of an antibody-immobilized QCM system detecting *Salmonella* spp. *Biosensors and Bioelectronics*, *15*, 167-172.

[59] X. L. Su and Y. B. Li (2005). A QCM immunosensor for *Salmonella* detection with simultaneous measurements of resonant frequency and motional resistance. *Biosensors and Bioelectronics*, *21*, 840-848.

[60] Z. H. Shen, M. C. Huang, C. D. Xiao, Y. Zhang, X. Q. Zeng and P. G. Wang (2007). Nonlabeled quartz crystal microbalance biosensor for bacterial detection using carbohydrate and lectin recognitions. *Analytical Chemistry*, *79*, 2312-2319.

[61] S. Huang, H. Yang, R. S. Lakshmanan, M. L. Johnson, I. Chen, J. Wan, H. C. Wikle, V. A. Petrenko, J. M. Barbaree, Z. Y. Cheng and B. A. Chin (2008). The Effect of Salt and Phage Concentrations on the Binding Sensitivity of Magnetoelastic Biosensors for *Bacillus anthracis* Detection. *Biotechnology and Bioengineering*, *101*, 1014-1021.

[62] R. Guntupalli, R. S. Lakshmanan, J. Hu, T. S. Huang, J. M. Barbaree, V. Vodyanoy and B. A. Chin (2007). Rapid and sensitive magnetoelastic biosensors for the detection of *Salmonella typhimurium* in a mixed microbial population. *Journal of Microbiological Methods*, *70*, 112-118.

[63] G. A. Campbell, J. Uknalis, S. I. Tu and R. Mutharasan (2007). Detect of *Escherichia coli* O157 : H7 in ground beef samples using piezoelectric excited millimeter-sized cantilever (PEMC) sensors. *Biosensors and Bioelectronics*, *22*, 1296-1302.

[64] G. A. Campbell and R. Mutharasan (2005). Detection of pathogen *Escherichia coli* O157 : H7 using self-excited PZT-glass microcantilevers. *Biosensors and Bioelectronics*, *21*, 462-473.

[65] G. A. Campbell and R. Mutharasan (2007). Method of Measuring *Bacillus anthracis* Spores in the Presence of Copious Amounts of *Bacillus thuringiensis* and *Bacillus cereus*. *Analytical Chemistry*, *79*, 1145-1152.

[66] A. P. Davila, J. Jang, A. K. Gupta, T. Walter, A. Aronson and R. Bashir (2007). Microresonator mass sensors for detection of *Bacillus anthracis* Sterne spores in air and water. *Biosensors and Bioelectronics*, *22*, 3028-3035.

[67] B. Dhayal, W. A. Henne, D. D. Doorneweerd, R. G. Reifenberger and P. S. Low (2006). Detection of *Bacillus subtilis* spores using peptide-functionalized cantilever arrays. *Journal of the American Chemical Society*, *128*, 3716-3721.

[68] D. Ramos, J. Tamayo, J. Mertens, M. Calleja and A. Zaballos (2006). Origin of the response of nanomechanical resonators to bacteria adsorption. *Journal of Applied Physics*, *100*, 106105.

[69] G. A. Campbell and R. Mutharasan (2008). Near real-time detection of *Cryptosporidium parvum* oocyst by IgM-functionalized piezoelectric-excited millimeter-sized cantilever biosensor. *Biosensors and Bioelectronics*, *23*, 1039-1045.

[70] R. A. Villamizar, A. Maroto, F. X. Rius, I. Inza and M. J. Figueras (2008). Fast detection of Salmonella Infantis with carbon nanotube field effect transistors. *Biosensors and Bioelectronics*, *24*, 279-283.

[71] P. A. Piunno, U. J. Krull, R. H. Hudson, M. J. Damha and H. Cohen (1995). Fiber-optic DNA sensor for fluorometric nucleic acid determination. *Analytical Chemistry*, *67*, 2635-2643.

[72] S. Hohng and T. Ha (2005). Single-molecule quantum-dot fluorescence resonance energy transfer. *Chemphyschem*, *6*, 956-960.

[73] J. R. E. Shepard (2006). Polychromatic microarrays: Simultaneous multicolor array hybridization of eight samples. *Analytical Chemistry*, *78*, 2478-2486.

[74] S. Tyagi and F. R. Kramer (1996). Molecular beacons: Probes that fluoresce upon hybridization. *Nature Biotechnology*, *14*, 303-308.

[75] R. Y. C. Kong, M. M. H. Mak and R. S. S. Wu (2009). DNA technologies for monitoring waterborne pathogens: A revolution in water pollution monitoring. *Ocean & Coastal Management*, *52*, 355-358.

[76] N. Y. Fortin, A. Mulchandani and W. Chen (2001). Use of real-time polymerase chain reaction and molecular beacons for the detection of *Escherichia coli* O157 : H7. *Analytical Biochemistry*, *289*, 281-288.

[77] M. Piliarik, L. Párová and J. Homola (2009). High-throughput SPR sensor for food safety. *Biosensors and Bioelectronics*, *24*, 1399-1404.

[78] A. J. Baeumner, R. N. Cohen, V. Miksic and J. H. Min (2003). RNA biosensor for the rapid detection of viable *Escherichia coli* in drinking water. *Biosensors and Bioelectronics*, *18*, 405-413.

[79] F. Farabullini, F. Lucarelli, I. Palchetti, G. Marrazza and M. Mascini (2006). Disposable electrochemical genosensor for the simultaneous analysis of different bacterial food contaminants. *Biosensors and Bioelectronics 22*, 1544-1549

[80] F. Azek, C. Grossiord, M. Joannes, B. Limoges and P. Brossier (2000). Hybridization assay at a disposable electrochemical biosensor for the attomole detection of amplified human cytomegalovirus DNA. *Analytical Biochemistry*, *284*, 107-113.

[81] B. Elsholz, R. Worl, L. Blohm, J. Albers, H. Feucht, T. Grunwald, B. Jurgen, T. Schweder and R. Hintsche (2006). Automated detection and quantitation of bacterial RNA by using electrical microarrays. *Analytical Chemistry*, *78*, 4794-4802.

[82] O. A. Loaiza, S. Campuzano, M. Pedrero, P. García and J. M. Pingarrón (2009). Ultrasensitive detection of coliforms by means of direct asymmetric PCR combined with disposable magnetic amperometric genosensors. *Analyst*, *134*, 34-37.

[83] R. H. Liu, J. N. Yang, R. Lenigk, J. Bonanno and P. Grodzinski (2004). Self-contained, fully integrated biochip for sample preparation, polymerase chain reaction amplification, and DNA microarray detection. *Analytical Chemistry*, *76*, 1824-1831.

[84] A. Lermo, E. Zacco, J. Barak, M. Delwiche, S. Campoy, J. Barbé, S. Alegret and M. I. Pividori (2008). Towards Q-PCR of pathogenic bacteria with improved electrochemical double-tagged genosensing detection. *Biosensors and Bioelectronics*, *23*, 1805-1811.

[85] W. C. Liao and J. A. A. Ho (2009). Attomole DNA Electrochemical Sensor for the Detection of *Escherichia coli* O157. *Analytical Chemistry*, *81*, 2470-2476.

[86] L. J. Wang, Q. J. Liu, Z. Y. Hu, Y. F. Zhang, C. S. Wu, M. Yang and P. Wang (2009). A novel electrochemical biosensor based on dynamic polymerase-extending hybridization for *E.coli* O157:H7 DNA detection. *Talanta*, *78*, 647-652.

[87] X. L. Mao, L. J. Yang, X. L. Su and Y. B. Li (2006). A nanoparticle amplification based quartz crystal microbalance DNA sensor for detection of *Escherichia coli* O157 : H7. *Biosensors and Bioelectronics*, *21*, 1178-1185.

[88] S.-H. Chen, V. C. H. Wu, Y.-C. Chuang and C.-S. Lin (2008). Using oligonucleotide-functionalized Au nanoparticles to rapidly detect foodborne pathogens on a piezoelectric biosensor. *Journal of Microbiological Methods*, 73, 7-17.

[89] V. C. H. Wu, S. H. Chen and C. S. Lin (2007). Real-time detection of *Escherichia coli* O157:H7 sequences using a circulating-flow system of quartz crystal microbalance. *Biosensors and Bioelectronics*, 22, 2967-2975.

[90] H. Sun, Y. Zhang and Y. Fung (2006). Flow analysis coupled with PQC/DNA biosensor for assay of *E. coli* based on detecting DNA products from PCR amplification. *Biosensors and Bioelectronics*, 22, 506-512.

[91] C. A. Rowe-Taitt, J. P. Golden, M. J. Feldstein, J. J. Cras, K. E. Hoffman and F. S. Ligler (2000). Array biosensor for detection of biohazards. *Biosensors and Bioelectronics*, 14, 785-794.

[92] M. Zayats, Y. Huang, R. Gill, C.-A. Ma and I. Willner (2006). Label-free and reagentless aptamer-based sensors for small molecules. *Journal of the American Chemical Society,*, 128, 13666-13667.

[93] T. O'Brien, L. H. Johnson, J. L. Aldrich, S. G. Allen, L. T. Liang, A. L. Plummer, S. J. Krak and A. A. Boiarski (2000). The development of immunoassays to four biological threat agents in a bidiffractive grating biosensor. *Biosensors and Bioelectronics*, 14, 815-828.

[94] W. J. Wilson, C. L. Strout, T. Z. DeSantis, J. Stilwell, A. V. Carrano and G. L. Andersen (2002). Sequence-specific identification of 18 pathogenic microorganisms using microarray technology. *Molecular and Cellular Probes*, 16, 119-127.

[95] S. M. Miller, D. M. Tourlousse, R. D. Stedtfeld, S. W. Baushke, A. B. Herzog, L. M. Wick, J. M. Rouillard, E. Gulari, J. M. Tiedje and S. A. Hashsham (2008). In situ-synthesized virulence and marker gene biochip for detection of bacterial pathogens in water. *Applied and Environmental Microbiology*, 74, 2200-2209.

[96] S. G. Bavykin, J. P. Akowski, V. M. Zakhariev, V. E. Barsky, A. N. Perov and A. D. Mirzabekov (2001). Portable system for microbial sample preparation and oligonucleotide microarray analysis. *Applied and Environmental Microbiology*, 67, 922-928.

[97] M. J. LaGier, J. W. Fell and K. D. Goodwin (2007). Electrochemical detection of harmful algae and other microbial contaminants in coastal waters using hand-held biosensors. *Marine Pollution Bulletin*, 54, 757-770.

[98] K. E. Sapsford, M. M. Ngundi, M. H. Moore, M. E. Lassman, L. C. Shriver-Lake, C. R. Taitt and F. S. Ligler (2006). Rapid detection of foodborne contaminants using an Array Biosensor. *Sensors and Actuators, B: Chemical*, *113*, 599-607.

[99] A. S. Mittelmann, E. Z. Ron and J. Rishpon (2002). Amperometric quantification of total coliforms and specific detection of *Escherichia coli*. *Analytical Chemistry*, *74*, 903-907.

[100] B. Serra, M. D. Morales, J. Zhang, A. J. Reviejo, E. H. Hall and J. M. Pingarron (2005). In-a-day electrochemical detection of coliforms in drinking water using a tyrosinase composite biosensor. *Analytical Chemistry*, *77*, 8115-8121.

[101] M. D. Morales, B. Serra, A. Guzman-Vazquez de Prada, A. J. Reviejo and J. M. Pingarron (2007). An electrochemical method for simultaneous detection and identification of *Escherichia coli*, *Staphylococcus aureus* and *Salmonella choleraesuis* using a glucose oxidase-peroxidase composite biosensor. *Analyst*, *132*, 572-578.

[102] C. A. Togo, V. C. Wutor, J. L. Limson and B. I. Pletschke (2007). Novel detection of *Escherichia coli* beta-D-glucuronidase activity using a microbially-modified glassy carbon electrode and its potential for faecal pollution monitoring. *Biotechnology Letters*, *29*, 531-537.

[103] F. G. Perez, M. Mascini, I. E. Tothill and A. P. Turner (1998). Immunomagnetic separation with mediated flow injection analysis amperometric detection of viable *Escherichia coli* O157. *Analytical Chemistry*, *70*, 2380-2386.

[104] J. D. Brewster and R. S. Mazenko (1998). Filtration capture and immunoelectrochemical detection for rapid assay of *Escherichia coli* O157:H7. *Journal of Immunological Methods*, *211*, 1-8.

[105] I. Abdel-Hamid, D. Ivnitski, P. Atanasov and E. Wilkins (1999). Flow-through immunofiltration assay system for rapid detection of *E. coli* O157:H7. *Biosensors and Bioelectronics*, *14*, 309-316.

[106] V. K. Rao, M. K. Sharma, A. K. Goel, L. Singh and K. Sekhar (2006). Amperometric immunosensor for the detection of *Vibrio cholerae* O1 using disposable screen-printed electrodes. *Analytical Science*, *22*, 1207-1211.

[107] Z. P. Aguilar and I. Fritsch (2003). Immobilized enzyme-linked DNA-hybridization assay with electrochemical detection for *Cryptosporidium parvum hsp70* mRNA. *Analytical Chemistry*, *75*, 3890-3897.

[108] F. Farabullini, F. Lucarelli, I. Palchetti, G. Marrazza and M. Mascini (2007). Disposable electrochemical genosensor for the simultaneous

analysis of different bacterial food contaminants. *Biosensors and Bioelectronics*, *22*, 1544-1549.

[109] C. Ruan, L. Yang and Y. Li (2002). Immunobiosensor chips for detection of Escherichia coli O157:H7 using electrochemical impedance spectroscopy. *Analytical Chemistry, 74,* 4814-4820.

[110] A. Shabani, M. Zourob, B. Allain, C. A. Marquette, M. F. Lawrence and R. Mandeville (2008). Bacteriophage-modified microarrays for the direct impedimetric detection of bacteria. *Analytical Chemistry*, *80*, 9475-9482.

[111] M. Zayats, O. A. Raitman, V. I. Chegel, A. B. Kharitonov and I. Willner (2002). Probing antigen-antibody binding processes by impedance measurements on ion-sensitive field-effect transistor devices and complementary surface plasmon resonance analyses: development of cholera toxin sensors. *Analytical Chemistry*, *74*, 4763-4773.

[112] A. G. Gehring, D. L. Patterson and S. I. Tu (1998). Use of a light-addressable potentiometric sensor for the detection of *Escherichia coli* O157:H7. *Analytical Biochemistry*, *258*, 293-298.

[113] C. A. Rowe-Taitt, J. P. Golden, M. J. Feldstein, J. J. Cras, K. E. Hoffman and F. S. Ligler (2000). Array biosensor for detection of biohazards. *Biosensors and Bioelectronics*, *14*, 785-794.

[114] P. Zhu, D. R. Shelton, J. S. Karns, A. Sundaram, S. Li, P. Amstutz and C. M. Tang (2005). Detection of water-borne *E. coli* O157 using the integrating waveguide biosensor. *Biosensors and Bioelectronics*, *21*, 678-683.

[115] S. D. Leskinen and D. V. Lim (2008). Rapid ultrafiltration concentration and biosensor detection of enterococci from large volumes of florida recreational water. *Applied and Environmental Microbiology*, *74*, 4792-4798.

[116] M. F. Kramer and D. V. Lim (2004). A rapid and automated fiber optic-based biosensor assay for the detection of *Salmonella* in spent irrigation water used in the sprouting of sprout seeds. *Journal of Food Protection*, *67*, 46-52.

[117] A. Almadidy, J. Watterson, P. A. E. Piunno, I. V. Foulds, P. A. Horgen and U. Krull (2003). A fibre-optic biosensor for detection of microbial contamination. *Canadian Journal of Chemistry-Revue Canadienne De Chimie*, *81*, 339-349.

[118] S. Ahn and D. R. Walt (2005). Detection of *Salmonella* spp. using microsphere-based, fiber-optic DNA microarrays. *Analytical Chemistry*, *77*, 5041-5047.

[119] C. K. Park, C. D. Kang and S. J. Sim (2008). Non-labeled detection of waterborne pathogen *Cryptosporidium parvum* using a polydiacetylene-based fluorescence chip. *Biotechnology Journal*, *3*, 687-693.

[120] B. H. Pyle, S. C. Broadaway and G. A. McFeters (1995). A rapid, direct method for enumerating respiring enterohemorrhagic *Escherichia coli* O157:H7 in water. *Applied and Environmental Microbiology*, *61*, 2614-2619.

[121] M. Magliulo, P. Simoni, M. Guardigli, E. Michelini, M. Luciani, R. Lelli and A. Roda (2007). A rapid multiplexed chemiluminescent immunoassay for the detection of *Escherichia coli* O157:H7, *Yersinia enterocolitica*, *Salmonella typhimurium*, and *Listeria monocytogenes* pathogen bacteria. *Journal of Agricultural Food and Chemistry*, *55*, 4933-4939.

[122] V. Koubova, E. Brynda, L. Karasova, J. Skvor, J. Homola, J. Dostalek, P. Tobiska and J. Rosicky (2001). Detection of foodborne pathogens using surface plasmon resonance biosensors. *Sensors and Actuators B-Chemical*, *74*, 100-105.

[123] C. Garcia-Aljaro, X. Munoz-Berbel, A. T. Jenkins, A. R. Blanch and F. X. Munoz (2008). Surface plasmon resonance assay for real-time monitoring of somatic coliphages in wastewaters. *Applied and Environmental Microbiology*, *74*, 4054-4058.

[124] J. B. Delehanty and F. S. Ligler (2002). A microarray immunoassay for simultaneous detection of proteins and bacteria. *Analytical Chemistry*, *74*, 5681-5687.

[125] K. E. Sapsford, A. Rasooly, C. R. Taitt and F. S. Ligler (2004). Detection of *Campylobacter* and *Shigella* species in food samples using an array biosensor. *Analytical Chemistry*, *76*, 433-440.

[126] C. R. Taitt, Y. S. Shubin, R. Angel and F. S. Ligler (2004). Detection of *Salmonella enterica* serovar typhimurium by using a rapid, array-based immunosensor. *Applied and Environmental Microbiology*, *70*, 152-158.

[127] Z. Wang, G. J. Vora and D. A. Stenger (2004). Detection and genotyping of *Entamoeba histolytica*, *Entamoeba dispar*, *Giardia lamblia*, and *Cryptosporidium parvum* by oligonucleotide microarray. *Journal of Clinical Microbiology*, *42*, 3262-3271.

[128] S. M. Miller, D. M. Tourlousse, R. D. Stedtfeld, S. W. Baushke, A. B. Herzog, L. M. Wick, J. M. Rouillard, E. Gulari, J. M. Tiedje and S. A. Hashsham (2008). In situ-synthesized virulence and marker gene biochip for detection of bacterial pathogens in water. *Applied and Environmental Microbiology*, *74*, 2200-2209.

[129] R. M. Carter, J. J. Mekalanos, M. B. Jacobs, G. J. Lubrano and G. G. Guilbault (1995). Quartz crystal microbalance detection of *Vibrio cholerae* O139 serotype. *Journal of Immunological Methods*, *187*, 121-125.
[130] M. Plomer, G. G. Guilbault and B. Hock (1992). Development of a piezoelectric immunosensor for the detection of enterobacteria. *Enzyme and Microbial Technology*, *14*, 230-235.
[131] S. T. Pathirana, J. Barbaree, B. A. Chin, M. G. Hartell, W. C. Neely and V. Vodyanoy (2000). Rapid and sensitive biosensor for *Salmonella*. *Biosensors and Bioelectronics*, *15*, 135-141.
[132] R. D. Vaughan, C. K. O'Sullivan and G. G. Guilbault (2001). Development of a quartz crystal microbalance (QCM) immunosensor for the detection of *Listeria monocytogenes*. *Enzyme and Microbial Technology*, *29*, 635-638.
[133] S. Tombelli, M. Mascini, C. Sacco and A. P. F. Turner (2000). A DNA piezoelectric biosensor assay coupled with a polymerase chain reaction for bacterial toxicity determination in environmental samples. *Analytica Chimica Acta*, *418*, 1-9.
[134] E. Howe and G. Harding (2000). A comparison of protocols for the optimisation of detection of bacteria using a surface acoustic wave (SAW) biosensor. *Biosensors and Bioelectronics*, *15*, 641-649.
[135] R. Guntupalli, R. S. Lakshmanan, J. Hu, T. S. Huang, J. M. Barbaree, V. Vodyanoy and B. A. Chin (2007). Rapid and sensitive magnetoelastic biosensors for the detection of *Salmonella typhimurium* in a mixed microbial population. *Journal of Microbiological Methods*, *70*, 112-118.
[136] R. S. Lakshmanan, R. Guntupalli, J. Hu, D. J. Kim, V. A. Petrenko, J. M. Barbaree and B. A. Chin (2007). Phage immobilized magnetoelastic sensor for the detection of *Salmonella typhimurium*. *Journal of Microbiological Methods*, *71*, 55-60.
[137] G. A. Campbell and R. Mutharasan (2008). Near real-time detection of *Cryptosporidium parvum* oocyst by IgM-functionalized piezoelectric-excited millimeter-sized cantilever biosensor. *Biosensors and Bioelectronics*, *23*, 1039-1045.

Chapter 5

DRAWBACKS AND FUTURE TRENDS OF PATHOGEN BIOSENSORS

The biosensor technology has experienced enormous advances since the first biosensor for electrochemical detection of glucose was reported in 1962 by Clark and Lyons [1]. Although the biosensor research attracts nowadays the interest of multidisciplinary researchers, the area of biomedicine condensed the majority of research on this field few decades ago. However, the need for rapid methods for bacterial detection in other areas such as food safety and clinical and environmental monitoring has attracted special interest for such devices. This is reflected in the increasing number of scientific papers released during the last decades, which account now for approximately 10% of the publications in the biosensor field. Nevertheless, in spite of the high number of publications, biosensors have not yet been successfully introduced in the market and only few equipments and devices are commercially available for bacterial detection [2]. The reasons are both technological and marketing related [3] and will be discussed in this section.

First of all, the sensitivity of biosensors must comply with the existing regulations on food safety in order to be accepted by producers and users. A reliable biosensor must be able to distinguish the target bacteria in a complex sample, especially when we talk about environmental or food samples, potentially containing multiple analytes and non-pathogenic bacteria. This can be a great challenge if we bear in mind that bacterial pathogens are found in-field in very low numbers, and frequently in viable but not cultivable states, what makes even more difficult their recovery [4]. Let's for example consider the limit of detection for coliforms in drinking water samples, which should ideally be below 1 CFU/100 mL [5]. Such numbers are still difficult to achieve with the currently existing biosensors if not coupled to selective pre-enrichment steps. In addition, biosensors are by now only compatible with the analysis of small volumes of sample (normally between 1 mL and few µL). This, although extremely important when small volume of sample is available, is not suitable for the majority of analyses in industry settings where analyses of large volumes are required. Thus, before reaching the market for environmental monitoring, biosensors will have to be improved to include automated pre-concentration treatment of the sample in order to endow with the analysis of large volumes and meet the law requirements. To be realistic, none of the current sensors would provide, at least not on real samples, as much sensitivity as the traditional plate culture technique [3]. Not forgetting that the high heterogeneity of a sample containing "big" particle-like bacteria into a complex matrix can give rise very easily to false negatives and interfere with a number of transduction formats.

Another reason is the high fabrication cost of the majority of the biosensors. This is normally the major drawback for industry managers to introduce a new product in the market. This may be surpassed by the development of cheaper sensing materials and by redirecting research towards the optimizstion of multiplexed sensors for the simultaneous detection of different bacteria and/or analytes. Also the reduction in size brought about by the micro-and-nanotechnology fabrication procedures will allow mass production with lower costs. As a result, lower production costs and reduction in the time required to conduct an assay would be reached making them more attractive.

The development of truly re-usable biosensors is another issue that needs to be addressed in order to reduce fabrication costs. Research is on progress to develop new biomaterials to be used as recognition elements displaying higher stability and longer lifespan than the currently existing receptors [6] An example of this are the biomimetic molecules such us molecular imprinted polymers and aptamers as substitute for antibodies and peptide nucleic acids as substitute for nucleic acid probes. Also, new immobilisation procedures that minimise the alterations in the recognition elements allowing homogeneous sensors and new sensing materials need being investigated.

Finally, it has to be noted that before one biosensor is accepted to enter the market, it has to be validated [7]. There are different guides for validation of quantitative methods. However, there is a complete lack of guidelines for the validation of qualitative methods. The diverse nature of biosensors and the fact that a high proportion of them are qualitative will make it difficult their approval in terms of the current legislation protocols.

In spite of the aforementioned drawbacks, the recent advances in micro and nanotechnology are placing biosensors in a very promising position. The incorporation of biofunctionalised nanomaterials as transducer elements has allowed sensitivity improvement up to single-cell detection in the presence of different sample matrices and also simultaneous detection of multiple microorganisms [8]. On the other hand, it is expected that nanomaterials utilisation will facilitate mass production of nanoscale size devices, resulting in the fabrication of low-cost biosensors. The development of "lab-on-a chip" devices is also contributing to revolutionise the field. Lab-on-chips are defined as integrated micro-devices with microfluidic channels and reservoirs, where every step needed for the pre-treatment and detection of a target microorganism can be performed. Used for multi–analyte detection, lab-on-chips will result in minimal operational requirements, working in a semi-automated way and also reducing the external signal noise, allowing in most cases pre-treatment of the sample on the same chip. On the other hand, the possibility to perform real-time analysis, a realistic option for most existing sensing formats, is expected to reduce economic loses to the manufacturers. Therefore, the biosensors are promising alternatives to some of the current detection methods, and once the major drawbacks are solved, they are expected to definitively reach the market industry.

REFERENCES

[1] L. C. Clark, Jr. and C. Lyons (1962). Electrode systems for continuous monitoring in cardiovascular surgery. *Annals of the New York Academy of Sciences*, *102*, 29-45.

[2] E. C. Alocilja and S. M. Radke (2003). Market analysis of biosensors for food safety. *Biosensors and Bioelectronics*, *18*, 841-846.

[3] D. Ivnitski, I. Abdel-Hamid, P. Atanasov and E. Wilkins (1999). Biosensors for detection of pathogenic bacteria. *Biosensors and Bioelectronics*, *14* 599-624.

[4] Y. Liu, A. Gilchrist, J. Zhang and X. F. Li (2008). Detection of viable but nonculturable *Escherichia coli* O157:H7 bacteria in drinking water and river water. *Applied and Environmental Microbiology*, *74*, 1502-1507.

[5] Anonymous (1991). Drinking water: National primary drinkingwater regulations; total coliform proposed rule. *Federal Register,54*, 24.

[6] L. S. Ferreira, M. B. De Souza, J. O. Trierweiler, O. Broxtermann, R. O. M. Folly and B. Hitzmann (2003). Aspects concerning the use of biosensors for process control: experimental and simulation investigations. *Computers and Chemical Engineering*, *27*, 1165-1173.

[7] S. Rodriguez-Mozaz, M. J. L. de Alda and D. Barcelo (2006). Biosensors as useful tools for environmental analysis and monitoring. *Analytical and Bioanalytical Chemistry*, *386*, 1025-1041.

[8] L. Yang and R. Bashir (2008). Electrical/electrochemical impedance for rapid detection of foodborne pathogenic bacteria. *Biotechnology Advances*, *26*, 135-150.

INDEX

A

absorption, 90, 91
absorption coefficient, 90
acceptor, 136
accessibility, 55, 68, 74
accidental, 7
acid, 30, 31, 34, 37, 39, 46, 47, 48, 49, 52, 60, 76, 84, 86, 105, 106, 113, 115, 124, 141, 142, 144, 157, 166
acoustic, 91, 95, 114, 118, 156, 163
acrylic acid, 113
acrylonitrile, 113
active site, 129
actuators, 155
acute, 8, 107
adenosine, 106
adenovirus, 149
adhesion, 88, 94, 138
adjustment, 135, 136
adsorption, 47, 56, 57, 73, 75, 84, 100, 134, 136, 157
aerobic, 124, 128
agar, 24, 25, 26, 29
agent, 14
agents, 7, 8, 11, 13, 20, 28, 33, 70, 71, 75, 103, 111, 113, 116, 153, 159
agglutination, 25
agglutination test, 25
agricultural, 6, 8
air, 95, 96, 137, 157
alcohol, 37, 98, 113
alcohol oxidase, 98, 113
alfalfa, 13
algae, 160
alkaline, 80, 129
alkaline phosphatase, 80, 129
alloys, 138
alpha, 127
alternative, 32, 39, 42, 65, 68, 74, 122, 125, 132, 142, 147
alternatives, 74, 167
aluminium, 146
amide, 64
amine, 41, 63, 65, 66, 71
amino, 28, 41, 59, 63, 64, 125, 127, 133
amino acid, 28
amino acids, 28
amino groups, 63, 133
ammonia, 130
amorphous, 138
amplitude, 82, 138
Amsterdam, 118
anaerobic, 128
analytical tools, 27, 143
aniline, 113
animals, 11, 14, 43, 45
annealing, 31
anorexia, 9
antibiotics, 5, 11, 46, 125

Antibodies, 39
antibody, 2, 29, 45, 98, 99, 100, 113, 117, 126, 130, 136, 148, 152, 153, 154, 156, 161
antigen, 41, 70, 145, 161
appetite, 9, 10
aqueous solution, 64, 95, 135, 137
aqueous solutions, 64, 95, 135
arginine, 106
assessment, 154
asymptomatic, 16
atomic force, 119
atomic force microscopy, 119
Atomic Force Microscopy, 96
ATP, 1, 122, 151
attachment, 112, 138, 147
Au nanoparticles, 159
availability, 40, 42

B

B. subtilis, 149
Bacillus, 108, 127, 128, 131, 137, 139, 142, 152, 156, 157
Bacillus subtilis, 128, 137, 157
Bacillus thuringiensis, 157
back, 92, 122, 142
back pain, 8
bacteria, xiii, 1, 2, 5, 7, 12, 24, 26, 27, 28, 29, 31, 34, 37, 38, 43, 44, 45, 46, 48, 57, 60, 73, 75, 76, 83, 84, 89, 91, 95, 97, 98, 100, 102, 111, 121, 123, 125, 127, 128, 130, 131, 132, 133, 134, 135, 136, 137, 138, 139, 142, 143, 144, 148, 149, 150, 151, 152, 154, 156, 157, 158, 161, 162, 163, 165, 166, 167, 168
bacterial, xiii, 5, 23, 24, 25, 27, 28, 34, 37, 38, 40, 43, 53, 73, 83, 94, 99, 100, 101, 102, 104, 112, 118, 122, 123, 125, 128, 134, 135, 136, 138, 139, 143, 147, 148, 149, 151, 155, 156, 158, 159, 161, 163, 165
bacterial cells, 53, 123, 138, 139
bacterial infection, 34

bacterial strains, 5
bacteriophage, 44, 106, 133, 136
bacteriophages, 44, 53, 101, 107, 135
Bacteriophages, 43, 133
bacterium, 14, 45
Badia, 115
base pair, 46, 48, 49, 142
battery, 28, 45
beams, 96
beef, 11, 117, 131, 153, 157
bending, 96, 139
binding, 39, 41, 42, 43, 44, 45, 46, 50, 52, 54, 55, 56, 57, 60, 63, 70, 71, 72, 73, 74, 76, 77, 81, 82, 83, 85, 88, 90, 92, 94, 106, 107, 111, 114, 121, 139, 148, 161
Bioanalytical, 98, 99, 102, 108, 112, 118, 119, 152, 168
biochemistry, 117
biocompatible, 38, 55
bioengineering, 37
biofilms, 107
biological interactions, 70
bioluminescence, 122, 151
biomaterials, 166
Biometals, iv
biomimetic, 54, 166
biomolecular, 104, 118
biomolecule, 37, 55, 56, 65, 66, 67, 76, 91
biomolecules, 67
bioreactors, 112
Biosensor, i, v, vii, 36, 37, 55, 77, 101, 160
biosensors, xiii, 35, 36, 37, 39, 47, 48, 53, 54, 57, 60, 73, 77, 80, 81, 82, 83, 84, 85, 86, 87, 89, 91, 93, 96, 97, 98, 102, 103, 104, 105, 110, 115, 117, 118, 119, 121, 122, 137, 138, 141, 145, 147, 154, 156, 160, 162, 163, 165, 166, 167, 168
biotechnological, 71
biotechnology, xiii, 113
bioterrorism, 33, 153

biotin, 39, 44, 46, 47, 48, 57, 60, 68, 70, 72, 73, 74, 101, 113, 114, 131, 144
blocks, 60
blood, 8, 25, 41
blot, 68
bonding, 56
bonds, 41, 56, 61, 62, 64
botulinum, 103
bovine, 75, 109, 114
brain, 10
brevis, 149
broad spectrum, 46
brucellosis, 11
buffer, 50, 133

C

campylobacter, 20
Campylobacter jejuni, 8, 135, 148, 149, 150
Canada, 16
cancer, 98
capacitance, 82, 151
capillary, 131, 153
capsule, 45
carbohydrate, 38, 45, 71, 73, 94, 102, 118, 138, 156
carbohydrates, 28, 84
carbon, 4, 28, 36, 39, 65, 77, 78, 109, 122, 126, 127, 133, 140, 151, 157, 160
carbon nanotubes, 78, 122, 140, 151
carboxyl, 41, 59, 63
carboxyl groups, 63
carboxylic, 133
carboxylic groups, 133
carrier, 57
casein, 75
catalase, 116, 128, 152
cattle, 11
cell, 25, 26, 29, 32, 38, 41, 43, 44, 53, 68, 73, 75, 77, 79, 87, 89, 90, 95, 122, 123, 127, 129, 134, 136, 139, 140, 147, 149, 167
cell culture, 25

cell line, 41
cell surface, 44, 68
cellulose, iv, 111
Centers for Disease Control, 15, 20, 21
chain molecules, 110
channels, 95, 135, 137, 167
charcoal, 25
charged particle, 84
cheese, 14
chemical interaction, 96, 119
chemical reactions, 71
children, 12
cholera, 9, 161
chromium, 88
classical, 23, 27, 50, 123, 129, 130, 135, 142
cloning, 42
Clostridium botulinum, 1, 9, 148
codes, 147
coding, 144, 147, 148
cohort, 18
coil, 138
coliforms, 28, 124, 128, 145, 151, 152, 158, 160, 165
colitis, 17
community, 16, 17, 20
competition, 81, 87, 149
complexity, 26, 82, 83
components, xiii, 29, 30, 31, 36, 37, 38, 42, 44, 47, 60, 63, 64, 65, 68, 73, 75, 76, 80, 82, 83, 84, 86, 92, 96, 121, 122, 123, 132, 133, 135, 136, 142, 147
composition, 38, 55, 58, 61, 70, 91, 133
compounds, 40, 63
concentration, 3, 25, 26, 29, 58, 82, 84, 85, 118, 123, 124, 126, 128, 129, 130, 131, 133, 137, 139, 143, 145, 146, 148, 156, 161, 166
conductance, 84, 133
conducting polymers, 112, 115
conduction, 91
conductive, 79, 104, 106
conductivity, 78, 82, 93, 124, 146
confinement, 67

congress, 35
Congress, viii
conjugation, 43, 44, 47, 57, 60, 63, 64, 65, 66, 68, 73, 137
construction, 47
consumers, 12
consumption, 7, 53, 81, 96, 125, 128
contaminant, 17
contaminants, 104, 149, 158, 160, 161
contamination, 11, 14, 20, 23, 32, 145, 162
control, 11, 60, 76, 79, 131, 134, 135, 139, 143
convection, 91
conversion, 124
cooking, 6, 14
copper, 58
correlation, 140
cost-effective, xiii
costs, 12, 36, 166
cough, 9
coupling, 88, 90, 111, 137
covalent, 39, 47, 55, 61, 62, 70
covalent bond, 55
covalent bonding, 55
crops, 11
crosslinking, 43
cross-linking, 51, 63, 133, 135
cryptosporidium, 15
culture, 2, 15, 23, 24, 25, 26, 27, 29, 43, 44, 84, 123, 126, 128, 133, 134, 137, 139, 166
culture conditions, 24
culture media, 24, 44, 84, 123, 126, 128, 133, 139
cycles, 26, 145
cyclic voltammetry, 146
cycling, 145
cyclohexane, 64
cysts, 16
cytomegalovirus, 115, 144, 158
cytometry, 27

D

dairy, 13, 14, 107, 116
dairy products, 13, 14
data analysis, 82, 83
death, 13, 14, 17
deaths, 5, 6, 12, 13
decomposition, 128
defects, 10
defence, 46
definition, 129
degradation, 47
dehydration, 95
denaturation, 31, 39, 57, 74
Denmark, 29
density, 39, 59, 93
deoxynivalenol, 149
deposition, 42, 44, 62, 67, 93, 132, 134
derivatives, 11, 36, 39, 64, 73, 81
detection, xiii, 6, 7, 15, 23, 24, 25, 26, 27, 28, 29, 30, 31, 32, 33, 34, 36, 38, 39, 40, 42, 45, 46, 47, 48, 49, 53, 54, 57, 60, 67, 68, 73, 75, 76, 77, 78, 80, 81, 82, 83, 84, 85, 86, 87, 89, 91, 92, 94, 95, 96, 97, 98, 99, 100, 101, 102, 103, 104, 105, 106, 107, 108, 109, 110, 111, 112, 114, 115, 116, 117, 118, 121, 122, 123, 125, 127, 128, 129, 130, 131, 132, 133, 134, 135, 136, 137, 138, 139, 140, 141, 142, 143, 144, 145, 146, 147, 148, 149, 150, 151, 152, 153, 154, 155, 156, 157, 158, 159, 160, 161, 162, 163, 164, 165, 166, 167, 168
detergents, 70, 75, 114
developing countries, 6, 37
diamines, 75
diarrhea, 17
diarrhoea, 8, 9, 13
dielectric constant, 88, 93
digestion, 41, 42
diodes, 86
discomfort, 8, 9, 10
discrimination, 143
diseases, 6, 7, 11, 19

displacement, 81, 86, 105
dissociation, 70, 72
distribution, 39, 139
diversity, 24
DNA, 1, 2, 4, 30, 31, 32, 35, 39, 46, 47, 48, 49, 76, 103, 104, 105, 106, 111, 112, 114, 115, 141, 142, 143, 144, 145, 146, 147, 148, 149, 157, 158, 159, 161, 162, 163
DNA polymerase, 31, 106
dogs, 14
donor, 136
dressings, 13
drinking, 6, 16, 17, 26, 33, 34, 126, 151, 158, 160, 165, 167
drinking water, 6, 16, 26, 33, 34, 126, 151, 158, 160, 165, 167
DuPont, 32
duration, 59

E

E. coli, 1, 4, 7, 8, 12, 13, 20, 21, 23, 24, 28, 48, 90, 94, 114, 121, 123, 124, 125, 126, 127, 128, 130, 131, 132, 133, 134, 135, 136, 137, 138, 139, 143, 144, 145, 146, 147, 149, 153, 159, 160, 161
eating, 11
E-coli, 111, 155
Eden, 17
egg, 73
elders, 12
electric field, 95
electrical properties, 82
Electroanalysis, 103, 104, 115
electrochemical detection, 80, 122, 129, 144, 145, 149, 151, 160, 161, 165
electrochemical impedance, 82, 97, 116, 161, 168
electrochemical measurements, 79
electrodes, 77, 78, 79, 80, 84, 127, 128, 132, 133, 134, 144, 145, 152, 161
electromotive force, 4

electron, 48, 60, 80, 82, 83, 88, 124, 130, 132
electrons, 80, 129
electrophoresis, 31
ELISA, 2, 25, 28, 29, 69, 78, 80, 114, 115, 129
emission, 85, 86, 142
employment, 142
endocarditis, 8
energy, 88, 122, 142, 157
energy transfer, 142, 157
enterococci, 161
enteroviruses, 26
entrapment, 47, 55, 66, 67, 73, 94, 112, 113, 138
environment, 6, 25, 27, 66
environment control, 27
environmental protection, 98
enzymatic, 30, 40, 41, 44, 47, 80, 81, 85, 152
enzymatic activity, 152
enzyme immunoassay, 115
enzyme-linked immunosorbent assay, 28, 108, 114, 153
enzymes, 1, 2, 28, 30, 37, 38, 44, 67, 80, 91, 112, 122, 127, 129, 132
epidemic, 18
epidemics, 6
epitope, 42
epitopes, 40, 55
equilibrium, 67, 82
Escherichia coli, 13, 17, 18, 42, 101, 103, 104, 109, 110, 116, 117, 131, 149, 150, 151, 152, 153, 154, 156, 157, 158, 159, 160, 161, 162, 167
estradiol, 109
ethanol, 58, 98
European Union, 2, 12, 19, 20
evolution, 50, 106, 152
excitation, 86, 117, 131
experimental condition, 88, 93
exploitation, 36, 122
exposure, 11, 31, 37, 53, 66, 140
extraction, 30, 54, 123
eye, 10, 26

F

fabrication, 36, 39, 49, 92, 166, 167
faecal, 23, 125, 135, 149, 160
false negative, 32, 86, 166
false positive, 32, 84, 148
fat, 13
fatigue, 10
FDA, 20, 21
Federal Register, 34, 167
fermentation, 28
ferromagnetic, 138
fertilizer, 11, 14
fever, 9
fiber, 85, 98, 117, 130, 131, 135, 136, 142, 153, 155, 161, 162
fibers, 136
film, 53, 67, 88, 92, 108, 113, 135
film formation, 67, 88
film thickness, 93
films, 67, 94, 103, 108, 110, 111, 112, 134, 154
filtration, 43, 126, 127, 128
fingerprinting, 28
fish, 11
FISH, 2, 30
FITC, 2
fixation, 31
flow, 79, 89, 116, 126, 139, 147, 149, 159, 160
fluorescein isothiocyanate (FITC), 142
fluorescence, 30, 31, 85, 123, 131, 136, 142, 148, 149, 157, 162
fluorescent microscopy, 31
fluorogenic, 28
fluorometric, 157
fluorophores, 2, 46, 86, 117, 132, 136, 142
focusing, xiii
folding, 50, 51, 52, 58, 64
food, xiii, 6, 10, 11, 12, 13, 15, 16, 23, 26, 27, 33, 34, 102, 104, 115, 116, 122, 147, 150, 151, 153, 158, 161, 162, 165, 167
Food and Drug Administration, 20, 21
food industry, xiii, 116, 151
food poisoning, 10, 11, 13
food products, 12
food safety, 27, 116, 158, 165, 167
foodborne illness, 12
forgetting, 7, 40, 166
fouling, 128
Fox, 15, 17
France, 27
free radical, 59
functionalization, 111
fusion, 101

G

garbage, 11
gas, 10
gastroenteritis, 8, 18
gastrointestinal, 17
gel, 111, 137
gene, 140, 144, 145, 146, 147, 149, 159, 163
generation, 50, 81, 127, 128
genes, 24, 47, 143, 144
genetic information, 140
genetics, 5
Geneva, 33
genome, 44, 73
genomic, 140
glass, 61, 86, 87, 88, 111, 131, 140, 149, 157
glucose, 37, 80, 98, 113, 125, 151, 160, 165
glucose oxidase, 80, 151, 160
glutaraldehyde, 63, 65
glycol, 75
glycoprotein, 45, 72, 73, 102
glycoproteins, 45
glycosylated, 73
gold, 36, 39, 43, 47, 51, 56, 58, 59, 77, 78, 88, 93, 99, 100, 108, 109, 110, 133, 135, 139, 144, 145, 147
gold nanoparticles, 147
gonadotropin, 115
graph, 89

graphite, 77, 126, 128
ground water, 129
groups, 24, 46, 47, 56, 58, 60, 61, 62, 63, 64, 65, 66, 71, 99, 110, 124, 133
growth, 24, 27, 53, 84, 93, 125, 128, 131, 133, 134, 151, 152, 154
growth rate, 154
guidelines, 33, 166

H

halogen, 135
halogens, 155
handling, 32, 65, 123, 126
hazards, 5
headache, 8, 9
health, 6, 12, 15, 27, 37
Heart, 125
heat, 31, 70, 91, 135, 137, 143, 150
heat loss, 91
heat shock protein, 143
Helicobacter pylori, 148
hemagglutinin, 45
hepatitis, 7, 9, 19
Hepatitis A, 9
hepatitis d, 19
herpes, 133, 154
herpes simplex, 133, 154
heterogeneity, 166
hip, 162
homogenized, 136
hormone, 115
hospitalization, 13
hospitalizations, 12, 14
hospitals, 5
host, 26, 44, 45, 46
House, 14, 21
household, 6
HRP, 2, 81, 129, 130, 145
human, xiii, 6, 13, 15, 66, 99, 109, 115, 144, 149, 158
human chorionic gonadotropin, 115
human subjects, 99
humans, 5, 6, 11, 148
hybrid, 142, 143

hybridization, 30, 31, 103, 105, 115, 157, 159, 161
hybridoma, 41
hybrids, 48, 144
hydro, 56, 75
hydrogels, 67
hydrogen, 56, 105, 128, 152
hydrogen peroxide, 128, 152
hydrolysis, 28, 123, 129
hydrolyzed, 51, 125
hydrophilic, 56, 75
hydrophobic, 56, 75
hydrophobic interactions, 56, 75
hydrophone, 138
hydroxyl, 60, 61, 65
hydroxyl groups, 61, 65
hygiene, 151

I

ice, 13
identification, 25, 28, 31, 55, 116, 124, 125, 149, 151, 152, 159, 160
IES, 82
IgE, 2, 69
IgG, 2, 40, 41, 68, 70, 114
Illinois, 113
illumination, 131
images, 79
imaging, 150
immersion, 62, 95, 96, 137
immobilization, 98, 99, 110, 111, 113
immortal, 41
immune response, 40, 46
immune system, 12
immunoassays, 107, 108, 113, 159
immunoglobulin, 140
Immunoglobulin E, 2
immunohistochemistry, 69
immunological, 25, 99, 113
immunoprecipitation, 69
impedance spectroscopy, 82, 104, 116, 132, 133, 161
implementation, 15, 107
IMS, 2, 29

Index

in situ, 30, 36, 74
in situ hybridization, 30
in vitro, 41, 42, 46, 50, 106
in vivo, 51
inactivation, 42, 150
incidence, 12, 88, 90
inclusion, 129, 143
incubation, 25, 28, 29, 33, 50, 59, 86, 94, 111, 125, 126, 127, 128, 131, 132, 133, 134, 135, 137, 138, 142, 145, 151
incubation period, 135
incubation time, 59, 133
indicators, 28, 135
indium, 36
indium tin oxide, 36
inducer, 2, 126, 127
induction, 125
industrial, 6
industry, xiii, 116, 151, 166, 167
inert, 38, 55, 67
infection, 12, 13, 15, 16, 26, 44, 45, 53, 133, 134, 136
infections, 12, 14, 18
infectious, 97, 103
infectious disease, 97
infectious diseases, 97
ingestion, 7
inhibitors, 30, 32
inhibitory, 32
injection, 44, 126, 137, 160
injury, viii
Innovation, iii
inoculation, 27, 40
inorganic, 66
instability, 31
instruments, 88, 134
insulin, 99
integration, 78, 81, 92
integrity, 38, 43, 53, 55, 67, 68
interaction, 56, 70, 74, 85
interactions, 47, 56, 70, 75, 94, 102, 104, 118, 119, 138, 156
interface, 48, 56, 77, 82, 88, 90, 124, 132, 136

interference, 29, 31, 125, 130, 131, 133, 136, 137
internal controls, 32
internalization, 12, 55
Internet, 154
ionic, 70, 75, 124
ions, 44, 84
iron, 138
irradiation, 91, 131
irrigation, 11, 161
ISO, 28
isoelectric point, 72, 73
isolation, 17, 24, 25, 26, 30, 33, 68, 145
isothermal, 144
Israel, 114

J

jaundice, 9
joint pain, 8

K

kidney, 14
kidney failure, 14
kinetics, 37
King, 98, 102, 118

L

label-free, 117, 153
lactic acid, 84, 124
lactic acid bacteria, 124
Lactobacillus, 128, 131
lactose, 24
Langmuir, 110, 138
laser, 27, 31, 86, 88, 90, 131, 138
law, 126, 166
lectin, 94, 102, 118, 138, 156
LED, 135, 155
Legionella, 9, 25, 33, 121, 135, 148, 155
Legionella pneumophila, 9, 25, 33, 135, 148, 155
legislation, 166

lettuce, 14
liberation, 44
lifespan, 166
ligand, 38, 54, 70, 91, 96, 107, 138
ligands, 45, 49, 87, 106
light beam, 89, 90, 142
limitation, 30, 53, 137, 139
limitations, 7, 27, 32, 74, 91, 143
linear, 94, 128, 138
lipid, 67, 112
lipopolysaccharide, 45
lipoproteins, 38
liposomes, 144, 145, 146
Listeria monocytogenes, 8, 14, 135, 136, 139, 144, 148, 150, 155, 162, 163
loading, 93, 94, 95
LOD, 2, 128, 137
London, 111
losses, 57
low back pain, 8
low power, 95
LPS, 3, 45
luciferin, 123
luminescence, 85
lymph, 10
lymph gland, 10
lysis, 26, 44, 53, 133, 135, 149

M

magnet, 130
magnetic, viii, 29, 35, 48, 124, 126, 130, 133, 138, 145, 158
magnetic beads, 145
magnetic field, 29, 138
magnetic particles, 124, 126, 130, 133
magnetoelastic, 101, 156, 163
malaise, 8, 9
mammal, 40
manipulation, xiii, 12, 14, 30, 31, 36, 43, 47, 126, 132, 148
manure, 11, 14
market, 47, 141, 165, 166, 167
marketing, 165
mass transfer, 83

matrix, 29, 66, 67, 75, 76, 77, 84, 130, 133, 145, 166
measurement, 83, 90, 95, 96, 107, 116, 132, 137, 139, 140, 148, 151
measures, 11, 88, 95
meat, 12, 13, 19, 138, 139
media, 24, 25, 28, 44, 64, 84, 90, 123, 126, 128, 132, 133, 135, 136, 139
mediators, 80, 129
melting, 47
membranes, 56, 146
meningitis, 8
mercury, 58
messenger RNA, 32
metabolic, 53, 123, 128
metabolism, 28, 83, 84, 122, 123, 128
metabolites, 53, 124
metal oxide, 61
metals, 58, 109, 117
methylene, 128, 152
metric, 1, 3
$MgSO_4$, 133
mica, 61
mice, 129, 131
microarray, 33, 45, 48, 49, 85, 102, 104, 111, 142, 144, 148, 158, 159, 160, 162, 163
microarray detection, 104, 149, 158
microarray technology, 33, 48, 148, 159
Microarrays, 86, 87, 106, 153
microbial, 19, 34, 54, 106, 107, 108, 115, 125, 151, 154, 156, 160, 162, 163
microelectrode, 82, 109, 151, 152, 153
microelectrodes, 78, 79, 132, 133
microfabrication, 36, 61, 92, 95
microfluidic channels, 95, 137, 167
microgels, 112
microorganism, 23, 25, 35, 38, 46, 76, 83, 122, 124, 128, 135, 142, 145, 149, 167
microorganisms, 5, 7, 11, 12, 23, 24, 27, 30, 32, 36, 39, 42, 45, 76, 85, 87, 89, 116, 122, 125, 128, 133, 138, 140, 145, 148, 159

micro-organisms, 167
micropatterning, 61
microscope, 30, 79
microscopy, 31, 119
microstructure, 95
microsystem, 35
milk, 11, 13, 14, 75, 128, 130, 133, 134, 137, 143, 150, 152, 154
miniaturization, 36, 95
MIP, 3, 54, 107
mirror, 50
Missouri, 17
mobility, 5
models, 24, 82
modulation, 138
mold, 124
mole, 3
molecular weight, 1, 70, 72, 108
molecules, 37, 40, 42, 46, 48, 50, 54, 55, 57, 58, 59, 61, 62, 63, 64, 66, 67, 68, 70, 71, 72, 73, 76, 88, 89, 103, 106, 108, 110, 124, 125, 130, 159, 166
molybdenum, 138
monoclonal, 38, 130, 137
monoclonal antibodies, 38, 130
monolayer, 26, 110, 111, 134, 138, 155, 156
monolayers, 43, 51, 57, 58, 98, 99, 100, 109, 110, 111
monomer, 55, 67, 72
monomeric, 54
monomers, 68
Monroe, 130
morphology, 93
motion, 119
mouse, 129
mouth, 9
movement, 80
mRNA, 3, 143, 161
multidisciplinary, 165
multiplicity, 134
muscle, 9, 10
muscle weakness, 9
mutations, 86
Mycobacterium, 127, 152

myoglobin, 67

N

nanomaterials, 167
nanoparticles, 48, 104, 133, 145, 147, 151, 153, 159
nanotechnology, 36, 109, 140, 166, 167
nanotube, 126, 157
nanowires, 122
National Academy of Sciences, 101
natural, 55, 149
nausea, 9, 10
New England, 15, 17
New York, vii, ix, 98, 118, 151, 167
New Zealand, 19
NHS, 3, 64, 65, 135, 137, 140
nickel, 138
Nielsen, 34, 104, 105
NIS, 3
nitride, 97, 139
noise, 79, 167
normal, 16
novel materials, 122
nuclease, 50
nucleic acid, 30, 31, 34, 37, 38, 39, 46, 47, 48, 49, 50, 52, 60, 71, 73, 76, 86, 105, 106, 115, 141, 142, 143, 144, 147, 157, 166
nylon, 128

O

oligonucleotides, 47, 104, 144
Oligosaccharides, v
on-line, 95, 137, 154
opposition, 2, 4, 40
optical, xiii, 35, 48, 49, 77, 85, 88, 90, 92, 115, 117, 135, 136, 142, 155
optical fiber, 135, 142, 155
optical properties, 88, 136
optimal performance, 38, 50, 95
optimization, 92
organic, 58, 66, 71, 108, 109, 124

organic solvent, 71
organic solvents, 71
organism, 45
orientation, 39, 63
oscillation, 91, 93
oxidation, 60, 65, 81, 122, 129
oxide, 146
oxygen, 1, 53, 61, 62, 128
oxygen consumption, 53

P

packaging, 14
pain, 8, 10
palladium, 58
paramagnetic, 29, 152
parameter, 134
parasite, 16
parasites, 12, 26
Parkinson, 97
particle-like, 166
particles, 29, 84, 124, 126, 130, 133, 142
passivation, 80
pasteurization, 11, 14
pathogenic, 7, 24, 34, 45, 97, 144, 145, 148, 156, 158, 159, 165, 167, 168
pathogens, 5, 6, 7, 12, 14, 15, 16, 19, 23, 24, 25, 33, 34, 47, 76, 100, 102, 106, 112, 115, 121, 124, 143, 145, 147, 148, 149, 153, 154, 155, 158, 159, 162, 163, 165
patients, 14
PCR, 3, 24, 30, 31, 32, 46, 48, 86, 114, 142, 143, 144, 145, 146, 147, 149, 158, 159
peptide, 44, 46, 49, 103, 105, 139, 141, 157, 166
Peptide, 3, 48, 105
peptides, 37, 39, 44, 46, 50, 57, 102, 103, 139
periodic, 7
permit, 133
peroxide, 128
perturbation, 82

phage, 1, 4, 42, 44, 53, 101, 126, 133, 135, 136, 152, 154
phenol, 126, 127
phosphate, 129, 144, 145
Phosphate, 3
photolithography, 58, 95
physical interaction, 44
physical properties, 35, 61, 77
physicochemical, 35
physico-chemical characteristics, 134
physics, 118
physiological, 73
piezoelectric, xiii, 35, 48, 77, 95, 99, 102, 110, 113, 140, 147, 154, 156, 157, 159, 163, 164
planar, 90
plants, 12, 45, 107, 128
plaque, 26
plaques, 26
plasma, 41, 99, 117
plasmons, 117, 155
plastic, 148
platforms, 29, 47, 95, 100, 152
platinum, 58, 77
PNA, 3, 48, 49, 104, 105, 106
pneumonia, 8
point mutation, 86
poisoning, 8, 10, 11, 13
polarization, 143
pollution, 107, 125, 135, 149, 158, 160
polyamide, 105
polyethylene, 75
polymer, 47, 54, 66, 67, 68, 75, 106
polymerase, 30, 32, 104, 146, 158, 159, 163
polymerase chain reaction, 30, 104, 158, 163
polymerisation conditions, 68
polymerization, 54, 55, 111
polymers, 54, 55, 76, 99, 112, 115, 166
polymethylmethacrylate, 137
polypeptide, 41
polysaccharide, 65
polysaccharides, 124
polystyrene, 28, 56, 114, 131

poor, 25, 36, 38, 57, 67
poor performance, 57
population, 5, 6, 23, 24, 154, 156, 163
pore, 25
pores, 133
pork, 11, 136
porous, 108, 113
poultry, 11, 13
power, 96
press, 111
prevention, xiii, 23
prions, 12
probe, 31, 33, 48, 76, 81, 100, 101, 103, 141, 142, 143, 147
process control, 98, 107, 168
producers, 165
production, 2, 4, 5, 15, 25, 36, 40, 42, 43, 44, 47, 50, 53, 54, 74, 78, 92, 95, 98, 128, 166, 167
production costs, 166
profit, 56, 78, 144
program, 82
proliferation, 26
propagation, 5, 17, 90, 91, 95
protection, 7, 98
protein, 13, 30, 39, 43, 50, 56, 66, 68, 69, 70, 72, 73, 75, 86, 98, 99, 100, 108, 113, 136, 138
protein arrays, 87
proteins, 38, 44, 45, 47, 49, 50, 56, 60, 67, 68, 69, 70, 72, 74, 108, 110, 111, 112, 114, 115, 124, 133, 162
proteomics, 87
protocol, 65
protocols, 36, 40, 93, 163, 166
protozoa, 7, 29, 140
protozoan, 16
pseudo, 107
Pseudomonas, 124, 130, 131, 148
Pseudomonas aeruginosa, 148
public, 12, 15
public health, 12, 15
pulse, 144, 149
pulses, 149
pumps, 145

purification, 68
pyruvate, 25

Q

quantum, 101, 142, 157
quantum dot, 142
quantum dots, 142
quartz, 88, 92, 93, 95, 99, 102, 104, 105, 118, 154, 156, 159, 163
query, 102

R

Raman, 119
random, 39, 42, 44, 55, 56, 57, 63, 73, 138
range, 5, 12, 29, 49, 55, 58, 71, 94, 123, 124, 126, 128, 134, 137, 138, 150, 155
rash, 9
reaction rate, 64
reactive groups, 63, 64
reactivity, 110, 138
reagent, 36, 92, 113
reagents, 32, 37, 71, 78, 94, 102
real time, 31, 36, 87, 88, 89, 132, 137, 156
receptors, 26, 39, 43, 44, 54, 56, 68, 166
recognition, 35, 36, 50, 54, 55, 70, 77, 94, 102, 103, 105, 133, 138, 150, 166
recovery, 50, 165
recreational, 6, 18, 26, 161
redox, 80, 129, 145
reflection, 90
refractive index, 85, 90, 135, 148
regenerate, 40, 74
regeneration, 50, 80
regular, 134
regulation, 11
regulations, 6, 27, 34, 130, 165, 167
relationship, 92
relevance, 12, 114
reliability, 135

replication, 133
reporters, 142
reputation, 13
reservoirs, 167
residues, 37
resistance, 20, 82, 83, 94, 118, 134, 156
resistive, 82
resonator, 91, 95
resources, 7
response time, 55, 136
restaurants, 20, 21
reusability, 48, 128
ribosomal, 31
rigidity, 140
risk, 6, 11, 15, 16, 31
risk factors, 11
RNA, 3, 30, 46, 48, 104, 105, 106, 144, 158
rotavirus, 7

S

safety, 19, 27, 116, 158, 165, 167
salad dressings, 13
saline, 133
salinity, 70
Salmonella, 7, 8, 9, 12, 13, 17, 20, 25, 34, 94, 99, 100, 101, 109, 118, 121, 124, 125, 128, 130, 131, 132, 133, 135, 136, 137, 138, 140, 144, 148, 149, 150, 151, 152, 153, 154, 155, 156, 157, 160, 161, 162, 163
salts, 49
sample, 26, 28, 29, 30, 31, 32, 36, 38, 53, 55, 70, 76, 77, 85, 86, 87, 90, 94, 96, 104, 121, 122, 123, 125, 126, 127, 129, 130, 132, 133, 137, 138, 139, 142, 145, 147, 148, 158, 160, 165, 167
sanitation, 11
scaffold, 99
Schmid, 98
search, 55, 56
seawater, 129, 143
sediments, 149

seeds, 162
selectivity, 25, 37, 42, 132, 141
Self, 3, 58, 98, 99, 104, 109, 110, 158
self-assembly, 51, 58, 60, 61, 62, 76, 110, 144
semicircle, 83
semiconductor, 142
semi-permeable membrane, 84
sensing, 35, 36, 38, 40, 47, 48, 53, 70, 73, 75, 76, 86, 89, 91, 92, 95, 108, 122, 123, 126, 131, 134, 135, 140, 152, 166, 167
sensitivity, 25, 29, 39, 40, 42, 47, 50, 55, 78, 92, 94, 101, 102, 105, 132, 136, 137, 139, 142, 144, 149, 152, 165, 167
sensors, 35, 44, 47, 48, 49, 53, 74, 75, 76, 80, 81, 85, 86, 90, 91, 92, 94, 95, 97, 107, 116, 117, 122, 130, 132, 134, 135, 136, 137, 138, 139, 143, 150, 154, 155, 157, 159, 161, 166
separation, 29, 86, 130, 145, 151, 160
series, 56, 145
serology, 102
serum, 70, 75, 109, 114
serum albumin, 70, 75, 114
services, viii
sewage, 128
shape, 24, 40, 58, 66, 74, 134, 142
Sheep, 69
shellfish, 11, 12
Shigella, 7, 8, 17, 18, 131, 162
shock, 143
shuttles, 129
signal transduction, 35, 40, 74, 76, 77, 80, 84, 88, 92, 132
signals, 89, 95, 127, 134, 138, 139
silane, 43, 61, 62, 63, 111
silica, 39, 43, 56, 61, 65, 67, 108, 111
silicon, 36, 61, 62, 89, 97, 133, 139, 154
silver, 58, 88
simulation, 110, 168
single cap, 36, 75
sites, 42, 54, 57
skills, 43

sludge, 8
sodium, 25, 65
software, 82
soil, 14, 128
solid state, 107
solvent, 58
solvents, 71
sorption, 118
spacers, 50, 51, 57
Spain, 18, 29
species, 7, 24, 38, 40, 50, 51, 68, 92, 97, 125, 131, 135, 140, 144, 150, 162
specific adsorption, 51, 60, 73, 75, 76, 92, 121, 132, 133, 135, 138
specificity, 29, 30, 32, 36, 37, 40, 42, 43, 46, 48, 66, 70, 75, 76, 81, 84, 92, 100, 122, 123, 126, 132, 135, 139
spectroscopy, 82, 104, 116, 132, 133, 161
spectrum, 11, 38, 46
speech, 9
spin, 137
spinach, 14
spore, 108, 139
SPR, 4, 38, 76, 85, 88, 89, 92, 101, 112, 122, 134, 135, 136, 143, 150, 154, 155, 158
sprouting, 5, 162
square wave, 145
stability, 39, 42, 49, 67, 141, 166
stabilizers, 13
standards, 30
staphylococcal, 113
Staphylococcus, 10, 68, 101, 125, 128, 138, 143, 144, 148, 149, 151, 160
Staphylococcus aureus, 10, 68, 101, 125, 128, 138, 143, 144, 148, 149, 151, 160
steady state, 128
steric, 57, 64, 86, 134
stomach, 10
storage, 15, 111, 151
strain, 13
strains, 5, 24, 68, 154

strategies, xiii, 6, 7, 15, 36, 37, 38, 57, 58, 60, 61, 67, 71, 92, 123, 137
streptavidin, 51, 70, 73, 99, 114, 130, 131, 142, 144, 145, 147
Streptomyces, 72, 73
stress, 15, 96, 139
structural changes, 108
structural protein, 38
substances, 32
substrates, 27, 42, 56, 61, 81, 84, 86, 110, 129, 132
success rate, 57
suffering, 66
sugar, 48
sulphate, 25
Sun, 114, 147, 159
supply, 6, 15, 16
supramolecular, 115
surface water, 16
surgery, 167
surveillance, 7, 23
surviving, 53
susceptibility, 28, 125
swallowing, 9
Sweden, 18
Switzerland, 18
symptoms, 8, 9, 10
syndrome, 8, 14
synthesis, 46, 145

T

targets, 31, 33, 39, 50, 57, 75, 86, 89, 94, 132, 134, 138, 145, 148, 149
teflon, 126, 128, 135
temperature, 59, 71, 79, 88, 93, 110, 135
thermal stability, 48
threat, 12, 113, 159
three-dimensional, 51, 58
thymine, 105
time frame, 128
tin, 36
tin oxide, 36
Togo, 160
total internal reflection, 90

Index

toxic, 40, 50, 66, 128
toxicity, 53, 107, 163
toxin, 25, 148, 161
toxins, 25, 30, 102, 103, 148, 150
trans, 91, 96, 132, 136, 140
transducer, 35, 38, 47, 77, 95, 109, 116, 130, 141, 142, 153, 167
transduction, 35, 36, 37, 49, 77, 80, 91, 92, 93, 130, 134, 141, 166
transfer, 48, 60, 80, 82, 83, 124, 130, 132, 142, 157
transistor, 161
transistors, 140, 157
transmission, 7, 8, 19, 23
transparent, 26, 90
trichinosis, 11
triggers, 40
tuberculosis, 11
typhoid, 8, 11
typhoid fever, 11
tyrosine, 74, 114

U

ultrasound, 137, 156
ultraviolet, 28
ultraviolet light, 28
uniform, 59
United States, 4, 18, 20, 101
urea, 130
urease, 130

V

vacuum, 93
validation, 122, 166
Van der Waals, 56
variation, 26, 82, 90, 135, 136, 140
vegetables, 11, 13, 14
vehicles, 14
velocity, 91, 95
versatility, 121
vibration, 88

Vibrio cholerae, 9, 109, 110, 129, 148, 152, 155, 161, 163
Vietnam, 19
viral gastroenteritis, 19
virulence, 24, 147, 148, 159, 163
virus, 18, 19, 26, 133, 154
virus infection, 26
viruses, 7, 9, 19, 26, 43, 45, 50, 108, 134
viscoelastic properties, 94
viscosity, 93
visible, 26, 88
vision, 9
voltammetric, 48, 68
vomiting, 9, 10

W

warfare, 116
waste disposal, 7, 78
waste water, 129
wastewater, 7, 53, 107
wastewater treatment, 7, 53, 107
wastewaters, 7, 107, 155, 162
water, 6, 7, 8, 14, 15, 16, 17, 18, 19, 23, 25, 26, 32, 33, 34, 107, 109, 115, 126, 128, 130, 134, 137, 148, 149, 151, 152, 153, 157, 158, 159, 160, 161, 162, 163, 165, 167
water quality, 6, 33
water resources, 7
waveguide, 90, 117, 130, 131, 136, 148, 153, 155, 161
wavelengths, 86
weakness, 9
weight loss, 10
wells, 11, 28, 78, 142
western blot, 68
whey, 13
wild animals, 11
wildlife, 11
workers, 123, 125, 127, 128, 129, 130, 131, 133, 134, 135, 136, 139, 143, 144, 145, 146, 147, 148, 149, 150
working conditions, 36
World Health Organization, 23, 33

Y

yeast, 5, 25, 124
yogurt, 150

Z

zoonosis, 12, 13
zoonotic, 11, 13, 20